Safety Sense

A Laboratory Guide

If you are exposed to toxic substances, or think you have been exposed to them, contact your local safety office immediately for instructions.

Emergency Phone Numbers

Safety Office:

Security Office:

Poison Control Center:

Institution Emergency:

Lab Emergency Contact:

Safety Sense
A Laboratory Guide

SECOND EDITION

http://www.cshprotocols.org

COLD SPRING HARBOR LABORATORY PRESS
Cold Spring Harbor, New York
http://www.cshlpress.com

Safety Sense: A Laboratory Guide
Second Edition

Publisher	John Inglis
Acquisition Editor	Jan Argentine
Technical Advisor	Maria Smit
Project Manager	Mary Cozza
Permissions Coordinator	Carol Brown
Production Editor	Patricia Barker
Desktop Editor	Susan Schaefer
Production Manager	Denise Weiss
Marketing Manager	Ingrid Benirschke

Front cover: Example of NFPA Chemical Hazard Label. **Health (Blue):** 4 **Danger** May be fatal on short exposure; 3 **Warning** Corrosive or Toxic; 2 **Warning** May be harmful if inhaled or absorbed; 1 **Caution** May be irritating; 0 No unusual hazard. **Flammability** (Red): 4 **Danger** Flammable gas or extremely flammable liquid; 3 **Warning** Flammable liquid flash point below 100°; 2 **Caution** Combustible liquid flash point of 100° to 200°; 1 Combustible if heated; 0 Not combustible. **Reactivity (Yellow):** 4 **Danger** Explosive material at room temperature; 3 **Danger** May be explosive if shocked, heated under confinement, or mixed with water; 2 **Warning** Unstable or may react violently if mixed with water; 1 **Caution** May react if heated or mixed with water but not violently; 0 **Stable** Not reactive when mixed with water. **Special Notice Key (White):** W Water reactive; OX Oxidizing agent.

Front inside cover: Periodic Table of the Elements (courtesy of Alfa® AESAR®, a Johnson Matthey Company)

Back inside cover: Commonly Used Hazard Symbols

Library of Congress Cataloging-in-Publication Data

Safety sense : a laboratory guide. -- 2nd ed.
  p. cm.
Rev. ed. of : Safety sense : a laboratory standard.
Includes bibliographical references and index.
ISBN 978-0-87969-783-9 (pbk. : alk. paper)
1. Hazardous substances--Handbooks, manuals, etc. 2. Industrial safety--Handbooks, manuals, etc. I. Title.

T55.3.H3S24 2007
660'.2804--dc22

2007002705

10 9 8 7 6 5 4 3 2 1

All Cold Spring Harbor Laboratory Press publications may be ordered directly from Cold Spring Harbor Laboratory Press, 500 Sunnyside Blvd., Woodbury, New York 11797-2924. Phone: 1-800-843-4388 in Continental U.S. and Canada. All other locations: (516) 422-4100. FAX: (516) 422-4097. E-mail: cshpress@cshl.edu. For a complete catalog of all Cold Spring Harbor Laboratory Press publications, visit our World Wide Web Site http://www.cshlpress.com/.

Contents

The material in this handbook was derived from the following Cold Spring Harbor Laboratory Press publications:

Antibodies: A Laboratory Manual
Arabidopsis: A Laboratory Manual
Archaea: A Laboratory Manual
Basic Methods in Microscopy
Bioinformatics: Sequence and Genome Analysis
Cells: A Laboratory Manual
Condensed Protocols From *Molecular Cloning: A Laboratory Manual*
DNA Microarrays: A *Molecular Cloning* Manual
DNA Science: A First Course in Recombinant DNA Technology
Drosophila: A Laboratory Manual
Drosophila: A Laboratory Handbook, 2nd edition
Drosophila Protocols
Early Development of *Xenopus laevis*: Course Manual
Eukaryotic Gene Expression
Experiments in Fission Yeast
Genome Analysis Laboratory Manual Series
Genetic Analysis of Pathogenic Bacteria: Course Manual
Gene Transfer: Delivery and Expression of DNA and RNA: A Laboratory Manual
Imaging Living Cells
Imaging Neurons: A Laboratory Manual
Imaging in Neuroscience & Development: A Laboratory Manual
Live Cell Imaging: A Laboratory Manual
Manipulating the Mouse Embryo: A Laboratory Manual, 2nd edition
Manipulating the Mouse Embryo: A Laboratory Manual, 3rd edition
Methods in Yeast Genetics
Methods in Yeast Genetics 2005 Edition. A Cold Spring Harbor Laboratory
    Course Manual
Molecular Cloning: A Laboratory Manual, 2nd edition
Molecular Cloning: A Laboratory Manual, 3rd edition
Molecular Probes of the Nervous System
Mouse Phenotypes: A Handbook of Mutation Analysis
PCR Primer: A Laboratory Manual
PCR Primer: A Laboratory Manual, 2nd edition
Phage Display: A Laboratory Manual
Protein–Protein Interactions: A *Molecular Cloning* Manual, 2nd edition
Proteins and Proteomics: A Laboratory Manual
Purifying Proteins for Proteomics: A Laboratory Manual
A Short Course in Bacterial Genetics
Strategies for Protein Purification and Characterization: A Laboratory Manual
Using Antibodies: A Laboratory Manual

Other Titles from CSHL Press

Lab Essentials

At the Bench: A Laboratory Navigator, Updated Edition

At the Helm: A Laboratory Navigator

Binding and Kinetics for Molecular Biologists

BioSupplyNet Laboratory Research Notebook

Experimental Design for Biologists

Lab Dynamics: Management Skills for Scientists

Lab Math: A Handbook of Measurements, Calculations, and Other Quantitative Skills for Use at the Bench

Lab Ref: A Handbook of Recipes, Reagents, and Other Reference Tools for Use at the Bench

Lab Ref, Volume 2: A Handbook of Recipes, Reagents, and Other Reference Tools for Use at the Bench

Introduction

A copy of a lab manual is not always readily available during the course of an experiment and safety information is not always at hand. Safety Sense: A Laboratory Guide, Second Edition is expanded and updated and is created from information contained in a wide range of Cold Spring Harbor Laboratory Press manuals past and present. The handbook makes important, general safety information easily accessible to laboratory workers. It is intended to be a practical, quick-reference guide, not a complete safety handbook. Its purpose is to maintain a safe and healthy research environment.

The primary safety information resources for laboratory personnel are

⊘ **Safety Office:** The best source of toxicity, hazard, storage, and disposal information is your local safety office, where a file of the most current information is maintained and available on request. Always consult this office for proper use and disposal procedures.

Obtain guidelines from your safety office for the appropriate gloves to be worn when handling each material, because disposable latex gloves may not afford the proper protection. For further information on gloves, see Appendix 4.

List the phone numbers for your local safety office, security office, poison control center, and lab emergency personnel on the first page of this publication and post them in an obvious place in your lab.

⊘ **Material Safety Data Sheets (MSDS):** The Occupational Safety and Health Administration (OSHA) requires that MSDS accompany all hazardous products that are shipped. These data sheets contain detailed safety information. MSDS should be filed in the laboratory in a central location as a reference guide.

It is essential for laboratory workers to be familiar with the potential hazards of materials used in laboratory experiments and to follow recommended procedures for their use, handling, storage, and disposal. Many icons have been adapted as an abbreviated warning for hazardous information, and a list of some of the international and/or commonly used hazard symbols is found on the inside back cover.

Other Reference Guides and Resources

A number of additional books and Web sites contain useful information about maintaining laboratory safety. The following are partial lists of these guides.

Reference Books

Alsimo R.J., ed. 2001. *Handbook of chemical health and safety*. American Chemical Society, Washington, D.C.

Barber V.C. and Clayton D.L., eds. 1985. *Electron microscopy safety handbook*. San Francisco Press.

Barker K. 1998. *At the bench: A laboratory navigator*. Cold Spring Harbor Laboratory Press, Cold Spring Harbor, New York.

Bevelacqua A.S. 2005. *Hazardous materials chemistry: Field operations guide*. Delmar Thomson Learning, Florence, Kentucky.

Biosafety in microbiological and biomedical laboratories (BMBL), 4th Edition. 1999. Centers for Disease Control, Washington, D.C. Online at http://www. cdc.gov/od/ohs/biosfty/bmbl4/bmbl4toc.htm

Chemical guide to the OSHA hazard communication standard, 8th Edition. 1994. Roytech Publications, Bethesda.

Davis J.M. 2002. *Basic cell culture: A practical approach*, 2nd Edition. Oxford University Press, New York.

Fleming D.O., Richardson J.H., Tuli J.I., and Vesley D., eds. 1995. *Laboratory safety: Principles and practices*, 2nd Edition. ASM Press, Washington, D.C.

Furr A.K. 2000. *CRC handbook of laboratory safety*, 5th Edition. 2000. CRC Press, Boca Raton, Florida.

Laboratory biosafety manual, 2nd Edition. 1993. World Health Organization, Geneva.

Laser safety guide, 10th Edition. 2000. Laser Institute of America, Orlando, Florida.

Lenga R.E., ed. 1988. *Sigma Aldrich library of chemical safety data*. Aldrich Chemical Company Library, St. Louis, Missouri.

Lewis R.J. 2000. *Rapid guide to hazardous chemicals in the workplace*, 4th Edition. Wiley-Interscience, New York.

Lewis R.J. 2002. *Hazardous chemicals desk reference*, 5th Edition. Wiley-Interscience, New York.

The Merck Index: An encyclopedia of chemicals, drugs, and biologicals.
2006. Wiley, New York.
NIOSH pocket guide to chemical hazards. 2005. National Institute for
Occupational Safety and Health, Washington, D.C. Online at
http://www. cdc.gov/niosh/npg
Plastics Design Library Staff. 2001. *Chemical resistance of plastics and
elastomers*, 3rd Edition. William Andrew Publishing/Plastics Design
Library, Norwich, New York.
Pohanish R.P. and Greene S.A. 2003. *Wiley guide to chemical incompatibilities*, 2nd Edition. Wiley-Interscience, New York.
Prudent practices in the laboratory: Handling and disposal of chemicals.
1998. National Research Council, Washington, D.C.

World Wide Web Sites

Useful information, databases, and links

Acronym Finder
    http://acronymfinder.com
Acronyms and Common Terms Used in Material Safety Data Sheets, Office
of Health and Safety, Centers for Disease Control and Prevention
    http://www.cdc.gov/od/ohs/manual/chemical/chmsaf5.htm
Agency for Toxic Substances and Disease Registry (ATSDR)
    http://www.atsdr.cdc.gov
American Biological Safety Association
    http://www.absa.org
American Chemical Society
    http://www.cas.org
American Type Culture Collection (ATCC)
    http://www.atcc.org
AnaSpec Chemical Abbreviations
    http://www.anaspec.com/html/chemical_abbreviations.html
Ansell Health Care, Chemical Resistance Glove Guide
    http://www.ansellpro.com/download/Ansell_7thEditionChemical
Resistance Guide.pdf
Biosafety in Microbiological and Biomedical Laboratories (BMBL), 4th
Edition, Centers for Disease Control and Prevention, Office of Health
and Safety
    http://www.cdc.gov/od/ohs/biosfty/bmbl4/bmbl4toc.htm

BioSupplyNet
  http://biosupplynet.com
CADS, MSDS Management
  http://www.cads.ca
Centers for Disease Control and Prevention, Office of Health and Safety,
  http://www.cdc.gov/od/ohs/hslinks.htm
ChemExper Chemical Directory
  http://www.chemexper.com
ChemFinder.Com, a portal of free and subscription scientific databases
  http://chemfinder.cambridgesoft.com
Chemical Reactivity Worksheet, NOAA, Office of Response and
    Restoration
  http://response.restoration.noaa.gov/index.php?keywords=chemical+
    reactivity+worksheet&submit=Go
Cold Spring Harbor Protocols
  http://www.cshprotocols.org
Cornell University Department of Environmental Health and Safety
    Center
  http://www.ehs.cornell.edu/msdssrch.asp
Environmental Health and Safety
  http://ehsfreeware.com/msdsinfo.htm
Genium's Group
  http://www.genium.com
GraphPad Online Calculators
  http://www.graphpad.com/quickalcs.index.cfm
Harvard University Biology Links
  http://mcb.harvard.edu/BioLinks.html
Hazardous Technical Information Services
  http://www.dscr.dla.mil/htis/htis.htm
Hazmat
  http://www.disasters.org/emgold/HAZMAT.htm
Howard Hughes Medical Institute Office of Laboratory Safety
  http://www.hhmi.org/research/labsafe
Institute of Occupational Safety Engineering (OSHWEB)
  http://www.oshweb.com
International Occupational Safety and Health Information Centre
  http://www.ilo.org/public/english/protection/safework/cis/products/
    safetytm/msds.htm
Internet Resources for MSDS
  http://www.ilpi.com/msds

Iowa State University, Department of Chemistry, MSDS
  http://avogadro.chem.iastate.edu/MSDS
Knovel
  http://www.knovel.com/knovel2/default.jsp
Los Alamos National Laboratory, Environmental Stewardship Division
  http://eweb.lanl.gov/Div_History.htm
Louisiana State University, CAMD
  http://www.camd.lsu.edu/msds/msds_search.html
Merck Chemical Database, ChemDAT Online
  http://www.chemdat.info/mda/index.html .
MSDS Hyper Glossary
  http://ilpi.com/msds/ref/index.html
MSDSonline
  http://www.msdsonline.com
MSDSSearch
  http://www.msdssearch.com
National Center for Biotechnology Information (NCBI), PubChem
  http://pubchem.ncbi.nlm.nih.gov
National Institute of Environmental Health and Human Services, Biological
    Safety
  http://www.niehs.nih.gov/odhsb/biosafe/bio.htm
National Institute of Environmental Health Sciences
  http://www.niehs.nih.gov
National Institute for Occupational Safety and Health, Database and
    Information Resources
  http://www.cdc.gov/niosh/database.html
National Institute for Occupational Safety and Health, Emergency Response
    Resources
  http://www.cdc.gov/niosh/topics/emres/ppe.html
National Institute of Standards and Technology, Standard Reference
    Materials
  http://ts.nist.gov/ts/htdocs/230/232/232.htm
National Institutes of Health
  http://www.nih.gov
National Library of Medicine TOXLINE, Bibliographic database for
    toxicology
  http://www.nlm.nih.gov/pubs/factsheets/toxlinfs.html
National Library of Medicine TOXNET, Databases on toxicology, hazardous
    chemicals, environmental health, and toxic releases
  http://toxnet.nlm.nih.gov

National Toxicology Program
  http://ntp-server.niehs.nih.gov
National Toxicology Program, National Institutes of Health, Report on
  Carcinogens
  http://ntp.niehs.nih.gov/ntp/roc/toc11.html
Occupational Safety and Health Administration (OSHA)
  http://www.osha.gov
Oklahoma State University, Environmental Health and Safety
  http://www.pp.okstate.edu/ehs/LINKS/msds.htm
Oxford University, Physical and Theoretical Chemistry Laboratory
  http://physchem.ox.ac.uk/MSDS/#MSDS
  http://www.pcl.ox.ac.uk/MSDS/abbreviations.html
Public Health Agency of Canada
  http://www.phac-aspc.gc.ca/msds-ftss/index.html
Seton Compliance Resource Center, MSDS/Hazard Communication Library
  http://www.setonresourcecenter.com/MSDSs/comply1.htm
Sigma-Aldrich, Reference material
  http://www.sigmaaldrich.com/Area_of_Interest /Equipment_Supplies_
  Books/Books/Reference
University of Akron, Chemical Database
  http://ull.chemistry.uakron.edu/erd
University of Chicago Libraries
  http://www.lib.uchicago.edu/e/su/chem/safetydata.html
University of Virginia, Laboratory Survival Manual
  http://keats.admin.virginia.edu/lsm/home.html
Vermont Safety Information Resources (SIRI)
  http://hazard.com

Storage, packaging, and shipping of biological materials

CDC Public Health Practice Program Office Shipping Guidelines
  http://www.phppo.cdc.gov/nltn/pdf/2000/3.0/shipdoc4.pdf
Occupational Health and Safety Administration, Department of Labor
  http://www.osha.gov
Seton Compliance Resource Center, Regulatory information
  http://www.setonresourcecenter.com
United States Department of Agriculture
  http://www.ocio.usda.gov/directives/doc/DR4400-007.htm
United States Department of Commerce
  http://www.commerce.gov

United States Department of Health and Human Services, Division of
  Occupational Health and Safety
  http://dohs.ors.od.nih.gov/shipping_biological_material_main.htm
United States Department of Health and Human Services, Policies and
  Regulations
  http://www.hhs.gov/policies/index.shtml

General Safety and Disposal Cautions

The guidance offered here is intended to be generally applicable. However, proper waste disposal procedures vary among institutions; therefore, always consult your local safety office for specific instructions. All chemically constituted waste must be disposed of in a suitable container clearly labeled with the type of material it contains and the date the waste was initiated.

The following general cautions should always be observed.

○ Become **completely familiar** with the properties of the substances **before** beginning the procedure.

○ The **absence of a warning** does not necessarily mean that the material is safe, since information may not always be complete or available.

○ If **exposed** to toxic substances, contact your local safety office immediately for instructions.

○ Use **proper disposal procedures** for all chemical, biological, and radioactive waste (see below, Disposal of Laboratory Waste).

○ For specific guidelines on **appropriate gloves**, consult your local safety office (also see Appendix 4).

○ Handle **concentrated acids and bases** with great care. Wear goggles and appropriate gloves. A face shield should be worn when handling large quantities.
   Do not mix strong acids with organic solvents because they may react. Sulfuric acid and nitric acid especially may react exothermically and cause fires and explosions.
   Do not mix strong bases with halogenated solvents because they may form reactive carbenes, which can lead to explosions.

○ Never **pipette** solutions using mouth suction. This method is not sterile and can be dangerous. Always use a pipette aid or bulb.

○ **Keep halogenated and nonhalogenated** solvents separately (e.g., mixing chloroform and acetone can cause unexpected reactions in the presence of bases). Halogenated solvents are organic solvents such as chloroform, dichloromethane, trichlorotrifluoroethane, and dichloroethane. Some nonhalogenated solvents are pentane, heptane, ethanol, methanol, benzene, toluene, *N,N*-dimethylformamide (DMF), dimethyl sulfoxide (DMSO), and acetonitrile.

⊘ **Laser radiation**, visible or invisible, can cause severe damage to the eyes and skin. Take proper precautions to prevent exposure to direct and reflected beams. Always follow manufacturers' safety guidelines and consult your local safety office. See caution below for more detailed information.

⊘ **Flash lamps**, due to their light intensity, can be harmful to the eyes. They also may explode on occasion. Wear appropriate eye protection and follow the manufacturer's guidelines.

⊘ Some materials used to clean **glassware** may be hazardous. Refer to the manufacturer's guidelines before using any solutions.

⊘ **Photographic fixatives and developers** also contain chemicals that can be harmful. Handle them with care and follow manufacturer's directions.

⊘ **Power supplies and electrophoresis equipment** pose serious fire hazards and electrical shock hazards if not used properly.

⊘ **Microwave ovens and autoclaves** in the lab require certain precautions. Accidents have occurred involving their use (e.g., to melt or sterilize agar stored in bottles). If the screw top is not completely removed and there is not enough space for the steam to vent, the bottles can explode and cause severe injury when the containers are removed from the microwave or autoclave. Always completely remove bottle caps before microwaving or autoclaving. An alternative method for routine agarose gels that do not require sterile agarose is to weigh out the agarose and place the solution in a flask.

⊘ Use extreme caution when handling **cutting devices** such as microtome blades, scalpels, razor blades, or needles. Microtome blades are extremely sharp! Use care when sectioning. If you are unfamiliar with their use, have someone demonstrate proper procedures. For proper disposal, use the "sharps" disposal container in your lab. Discard used needles unshielded, with the syringe still attached. This prevents injuries (and possible infections; see below, Biological Safety Procedures) while manipulating used needles, since many accidents occur while trying to replace the needle shield. Injuries may also be caused by broken Pasteur pipettes, coverslips, or slides.

Disposal of Laboratory Waste

There are specific regulatory requirements for the disposal of all medical waste and biological samples mandated by the U.S. Environmental Protection Agency and regulated by the individual states and territories. Medical and biological samples that require special handling and disposal are generally termed Medical Pathological Waste (MPW), and medical, veterinary, and biological facilities will have programs for the collection of MPW and its disposal. Restrictions on how radioactive waste can be disposed of are regulated by the U.S. Nuclear Regulatory Commission in 10 CFR 20.2001, General requirements for waste disposal, or the individual Agreement States. The preferred method for the disposal of radioactively contaminated MPW is decay-in-storage.

Waste and any materials contaminated with biohazardous materials must be decontaminated and disposed of as regulated medical waste. No harmful substances should be released into the environment in an uncontrolled manner. This includes all tissue samples, needles, syringes, scalpels, etc. Be sure to contact your institution's safety office concerning the proper practices associated with the handling and disposal of biohazardous waste.

Some basic rules are outlined below. For treatment of radioactive and biological waste, see sections below on Radioactive Safety Procedures and Biological Safety Procedures.

- In practice, only **neutral aqueous solutions** without heavy metal ions and without organic solvents can be poured down the drain (e.g., most buffers). Acid and basic aqueous solutions need to be neutralized cautiously before their disposal by this method.

- For proper disposal of **strong acids and bases**, dilute them by placing the acid or base onto ice, and neutralize them. Do not pour water into them. If the solution does not contain any other toxic compound, the salts can be flushed down the drain.

- For disposal of **other liquid waste**, similar chemicals can be collected and disposed of together, whereas chemically different wastes should be collected separately. This avoids chemical reactions between components of the mixture (see above). Collect at least inorganic aqueous waste, nonhalogenated solvents, and halogenated solvents separately.

- Waste from **photo processing and automatic developers** should be collected separately in order to recycle the silver traces found in it.

Radioactive Safety Procedures

In the USA and other countries, the access to radioactive substances is strictly controlled. You may be required to become a registered user (e.g., by attending a mandatory seminar and receiving a personal dosimeter). For properties, decay tables, and special considerations of radiochemicals, see Appendix 2. A convenient calculator to perform routine radioactivity calculations can also be found at:

> http://www.graphpad.com/calculators/radcalc.cmf

If you have never worked with radioactivity before, follow the steps below:

- **Try to avoid it!** Many experiments that are traditionally performed with the help of radioactivity can now be carried out using alternatives based on fluorescence or chemiluminescence, including, for example, DNA sequencing, Southern and Northern blots, and protein kinase assays.

- **Be informed.** While planning an experiment that involves the use of radioactivity, consider the physico-chemical properties of the isotope (half-life, emission type and energy), the chemical form of the radioactivity, its radioactive concentration (specific activity), total amount, and its chemical concentration. Order and use only as much as is really needed.

- **Familiarize yourself** with the designated working area. Perform a mental and practical dry run (replacing radioactivity by a colored solution) to make sure that all equipment needed is available and to get used to working behind a shield. Handle your samples as if sterility would be required to avoid contamination.

- **Always wear appropriate gloves**, lab coat, and safety goggles when handling radioactive material (see Appendix 4).

- **Check the work area** for contamination before, during, and after your experiment (including your lab coat, hands, and shoes).

- **Localize your radioactivity.** Avoid formation of aerosols or contamination of large volumes of buffers.

- **Liquid scintillation cocktails** are often used to quantitate radioactivity. They contain organic solvents and small amounts of organic compounds. Try to avoid contact with the skin. After use, they should be regarded as radioactive waste; the filled vials are usually collected in designated containers, separate from other (aqueous) liquid radioactive waste.

- **Dispose of radioactive waste** only into designated, shielded containers (separated by isotope, physical form [dry/liquid], and chemical form [aqueous/organic solvent phase]). Always consult your safety office for further guidance in the appropriate disposal of radioactive materials.

- Among the experiments requiring **special precautions** are those that use [^{35}S]methionine and ^{125}I, due to the dangers of air-borne radioactivity. [^{35}S]methionine decomposes during storage into sulfoxide gases, which are released when the vial is opened. The isotope ^{125}I accumulates in the thyroid and is a potential health hazard. ^{125}I is used for the preparation of Bolton-Hunter reagent to radioiodinate proteins. Consult your local safety office for further guidance in the appropriate use and disposal of these radioactive materials prior to initiating any experiments. Wear appropriate gloves when handling potentially volatile radioactive substances, and work only in a radioiodine fume hood.

Biological Safety Procedures

Biological safety fulfills three purposes: To avoid contamination of your biological sample with other species; to avoid exposure of the researcher to the sample; and to avoid release of living material into the environment. Biological safety begins with the receipt of the living sample; continues with its storage, handling, and propagation; and ends only with the proper disposal of all contaminated materials. A catalog of operations known as "sterile handling" is usually employed in manipulating living matter. However, the actual manner of treatment depends largely on the actual sample, which can be quite diverse: *E. coli* and other bacterial strains, yeasts, tissues of animal or plant origin, cultures of mammalian cells, or even derivatives from human blood are routinely handled in a biological laboratory. Two of these, bacteria and human blood products, are discussed in more detail below.

The Department of Health, Education, and Welfare (HEW) has classified various bacteria into different categories with regard to shipping requirements (see Sanderson and Zeigler, *Methods Enzymol.* 204: 248–264 [1991]). Nonpathogenic strains of *E. coli* (such as K12) and *B. subtilis* are in Class 1 and are considered to present no or minimal hazard under normal shipping conditions. However, *Salmonella, Haemophilus*, and certain strains of *Streptomyces* and *Pseudomonas* are in Class 2. Class 2 bacteria are

"Agents of ordinary potential hazard: agents which produce disease of varying degrees of severity... but which are contained by ordinary laboratory techniques." For further information regarding shipping, see Other Reference Guides and Resources section of this book on page 2.

Human blood, blood products, and tissues may contain occult infectious materials such as hepatitis B virus and HIV that may result in laboratory-acquired infections. Investigators working with lymphoblast cell lines transformed by Epstein-Barr virus (EBV) are also at risk of EBV infection. Any human blood, blood products, or tissues should be considered a biohazard and should be handled accordingly until proved otherwise. Wear appropriate disposable gloves, use mechanical pipetting devices, work in a biological safety cabinet, protect against the possibility of aerosol generation, and disinfect all waste materials before disposal. Autoclave contaminated plasticware before disposal; autoclave contaminated liquids or treat with bleach (10% [v/v] final concentration) for at least 30 minutes before disposal (this is valid also for used bacterial media).

Always consult your local institutional safety officer for specific handling and disposal procedures of your samples. Further information can be found in the Frequently Asked Questions of the ATCC homepage (http://www.atcc.org) or in the books *At the bench: A laboratory navigator* by K. Barker (1998) and *Basic cell culture: A practical approach*, 2nd Edition, edited by J.M. Davis (2002). Further information is also available from the National Institute of Environmental Health and Human Services, Biological Safety (http://www.niehs.nih.gov/odhsb/biosafe/bio.htm).

General Properties of Common Chemicals

The hazardous materials list can be subdivided into the following categories:

- Inorganic acids, such as hydrochloric, sulfuric, nitric, or phosphoric are colorless liquids with stinging vapors. Avoid spills on skin or clothing. Spills should be diluted with large amounts of water. The concentrated forms of these acids can destroy paper, textiles, and skin, as well as cause serious injury to the eyes.

- Inorganic bases such as sodium hydroxide are white solids that dissolve in water and under heat development. Concentrated solutions will slowly dissolve skin and even fingernails.

- Salts of heavy metals are usually colored, powdered solids that dissolve in water. Many of them are potent enzyme inhibitors and therefore toxic to humans and to the environment (e.g., fish and algae).

- Most organic solvents are flammable volatile liquids. Avoid breathing the vapors, which can cause nausea or dizziness. Also avoid skin contact.

- Other organic compounds, including organosulfur compounds such as mercaptoethanol or organic amines, can have very unpleasant odors. Others are highly reactive and should be handled with appropriate care.

- If improperly handled, dyes and their solutions can stain not only your sample, but also your skin and clothing. Some of them are also mutagenic (e.g., ethidium bromide), carcinogenic, and toxic.

- All names ending with "ase" (e.g., catalase, β-glucuronidase, zymolyase) refer to enzymes. There are also other enzymes with non-systematic names like pepsin. Many of them are provided by manufacturers in preparations containing buffering substances, etc. Be aware of the individual properties of materials contained in these substances.

- Toxic compounds are often used to manipulate cells. They can be dangerous and should be handled appropriately.

- Incompatible chemicals should be kept segregated (see Appendix 1).

- For information on resistance of some plastics to chemicals and sterilization, see Appendix 3.

- Be aware that several of the compounds listed have not been thoroughly studied with respect to their toxicological properties. Handle each chemical with the appropriate respect. Although the relative effects of toxic compounds can be quantified (e.g., LD_{50} values), this is not possible for carcinogens or mutagens, where one single exposure can have an effect. Be aware that dangers related to a given compound may also depend on its physical state (e.g., dry ice vs. carbon dioxide under pressure in a gas bomb). Anticipate under which circumstances during an experiment exposure is most likely to occur and how best to protect yourself and your environment.

Cold Spring Harbor Laboratory Press (CSHLP) has used its best efforts in collecting and preparing the material contained herein but does not assume, and hereby disclaims, any liability for any loss or damage caused by errors and omissions in the publication, whether such errors and omissions result from negligence, accident, or any other cause. CSHLP does not assume responsibility for the user's failure to consult more complete information regarding the hazardous substances listed in this publication and does not warrant this handbook to be a complete listing of hazardous substances, but expressly states that it is a listing of cautions about hazardous materials that can be found in various CSHLP laboratory manuals.

Hazardous Materials

ABTS, *see* **2,2´-Azino-bis(3-ethylbenzthiazoline)-6-sulfonic acid**

Acacia powder, *see* **Gum arabic**

Acetaminophen is harmful by inhalation, ingestion, or skin absorption. Wear appropriate gloves and safety goggles and always use in a chemical fume hood. Do not breathe the dust.

Acetic acid (concentrated) must be handled with great care. It may be harmful by inhalation, ingestion, or skin absorption. Wear appropriate gloves and goggles. Use in a chemical fume hood

Acetic acid (glacial) is highly corrosive and must be handled with great care. It may be a carcinogen. Liquid and mist cause severe burns to all body tissues. It may be harmful by inhalation, ingestion, or skin absorption. Wear appropriate gloves and goggles and use in a chemical fume hood. Keep away from heat, sparks, and open flame.

Acetic anhydride is extremely destructive to the skin, eyes, mucous membranes, and upper respiratory tract. It may be harmful by inhalation, ingestion, or skin absorption. Wear appropriate gloves and safety glasses and use in a chemical fume hood.

Acetone causes eye and skin irritation and is irritating to mucous membranes and upper respiratory tract. Do not breathe the vapors. It is also extremely flammable. Wear appropriate gloves and safety glasses. Keep away from heat, sparks, and open flame.

Acetonitrile (Methyl cyanide) is very volatile and extremely flammable. It is an irritant and a chemical asphyxiant that can exert its effects by inhalation, ingestion, or skin absorption. Treat cases of severe exposure as cyanide poisoning. Wear appropriate gloves and safety glasses and use only in a chemical fume hood. Keep away from heat, sparks, and open flame.

Acetylcholine causes severe irritation to eyes, skin, and respiratory tract. It may be harmful by inhalation, ingestion, or skin absorption. Wear appropriate gloves and safety glasses. Do not breathe the dust.

Acetylthiocholine iodide may be harmful by inhalation, ingestion, or skin absorption. Wear appropriate gloves and safety glasses and use in a chemical fume hood.

Acridine orange may be a mutagen and may be harmful by inhalation, ingestion, or skin absorption. Wear appropriate gloves and safety glasses. Do not breathe the dust.

Acrolein is extremely toxic and volatile. It may be harmful by inhalation, ingestion, or skin absorption. Wear appropriate gloves and safety goggles and use in a chemical fume hood. Keep away from heat, sparks, and open flame.

Acrylamide (unpolymerized) is a potent neurotoxin and is absorbed through the skin (the effects are cumulative). Avoid breathing the dust. Wear appropriate gloves and a face mask when weighing powdered acrylamide and methylene-bisacrylamide. Use in a chemical fume hood. Polyacrylamide is considered to be nontoxic, but it should be handled with care because it might contain small quantities of unpolymerized acrylamide.

Actinomycin D is a teratogen and a carcinogen. It is highly toxic and may be fatal if inhaled, ingested, or absorbed through the skin. It may also cause irritation. Avoid breathing the dust. Wear appropriate gloves and safety glasses and always use in a chemical fume hood. Solutions of actinomycin D are light-sensitive.

Adenosine is an irritant and may be harmful by inhalation, ingestion, or skin absorption. Wear appropriate gloves and safety glasses and use in a chemical fume hood. Do not breathe the dust.

Adenosine 3′, 5′-cyclic monophosphate (cAMP) is an irritant and is flammable. It contains a radioactive isotope that may produce cancer and genetic mutation. It is harmful by inhalation, ingestion, or skin absorption. Wear appropriate gloves and safety goggles and use in a chemical fume hood.

S-Adenosyl methionine (SAM) is toxic and may be harmful by inhalation, ingestion, or skin absorption. Wear appropriate gloves and safety glasses and use in a chemical fume hood. Do not breathe the dust.

AEBSF, *see* 4-(2-Aminoethyl)benzenesulfonyl fluoride hydrochloride

AEC, *see* Aminoethylcarbazole

AgNO$_3$, *see* Silver nitrate

Alanine, *see* L-DOPA

Aldehyde, *see* Formaldehyde

Aluminum chloride is corrosive and causes burns. It reacts violently with water. It may be harmful by inhalation, ingestion, or skin absorption. Wear appropriate gloves and safety goggles and use only in a chemical fume hood.

Aluminum fluoride may be harmful by inhalation, ingestion, or skin absorption. Wear appropriate gloves and safety glasses.

α-Amanitin is highly toxic and may be fatal by inhalation, ingestion, or skin absorption. Symptoms may be delayed for as long as 6–24 hours. Wear appropriate gloves and safety glasses and always use in a chemical fume hood.

Aminobenzoic acid may be harmful by inhalation, ingestion, or skin absorption. Wear appropriate gloves and safety glasses.

4-(2-Aminoethyl)benzenesulfonyl fluoride hydrochloride (AEBSF) is corrosive and may be harmful by inhalation, ingestion, or skin absorption. Wear appropriate gloves and safety glasses.

Aminoethylcarbazole (AEC) may be harmful by inhalation, ingestion, or skin absorption. Wear appropriate gloves and safety glasses.

ε-Amino-*n*-caproic acid (6-Aminocaproic acid), *see* Hexanoic acid

γ-Amino-*n*-butyric acid (GABA), *see* Butyric acid

3-Aminopropyltriethoxysilane (TESPA), *see* Silane

3-Amino-1,2,4-triazole (ATA) is a carcinogen. It may be harmful by inhalation, ingestion, or skin absorption. Wear appropriate gloves, safety glasses, and other protective clothing. Avoid breathing vapors. Use only in a chemical fume hood.

Ammonia, NH₃, is corrosive, toxic, and can be explosive. It may be harmful by inhalation, ingestion, or skin absorption. Use only with mechanical exhaust. Wear appropriate gloves and safety glasses.

Ammonium acetate, NH_4Ac, $H_3CCOONH_4$, may be harmful by inhalation, ingestion, or skin absorption. Wear appropriate gloves and safety glasses and use in a chemical fume hood.

Ammonium bicarbonate, NH_4HCO_3, may be harmful by inhalation, ingestion, or skin absorption. Wear appropriate gloves and safety glasses and use in a chemical fume hood.

Ammonium carbonate, $(NH_4)_2CO_3$, may be harmful by inhalation, ingestion, or skin absorption. Wear appropriate gloves and safety glasses and use in a chemical fume hood.

Ammonium chloride, NH_4Cl, may be harmful by inhalation, ingestion, or skin absorption. Wear appropriate gloves and safety glasses and use in a chemical fume hood.

Ammonium formate, *see* **Formic acid**

Ammonium hydrogen carbonate, *see* **Ammonium bicarbonate**

Ammonium hydroxide, NH_4OH, is a solution of ammonia in water. It is caustic and should be handled with great care. As ammonia vapors escape from the solution, they are corrosive, toxic, and can be explosive. Use only with mechanical exhaust. Wear appropriate gloves and use only in a chemical fume hood.

Ammonium molybdate, $(NH_4)_6Mo_7O_{24} \cdot 4H_2O$, (or its **tetrahydrate**), may be harmful by inhalation, ingestion, or skin absorption. Wear appropriate gloves and safety glasses and use in a chemical fume hood.

Ammonium nitrate, NH_4NO_3, may be harmful by inhalation, ingestion, or skin absorption. Wear appropriate gloves and safety glasses and use in a chemical fume hood. To avoid explosions, keep away from heat, sparks, and open flame.

Ammonium persulfate, $(NH_4)_2S_2O_8$, is extremely destructive to tissue of the mucous membranes and upper respiratory tract, eyes, and skin. Inhalation may be fatal. Wear appropriate gloves, safety glasses, and protective clothing, and use only in a chemical fume hood. Wash thoroughly after handling.

Ammonium phosphate, *see* **Phosphoric acid**

Ammonium sulfate, $(NH_4)_2SO_4$, may be harmful by inhalation, ingestion, or skin absorption. Wear appropriate gloves and safety glasses.

Ammonium sulfide causes severe irritation to the respiratory tract and may be harmful by inhalation, ingestion, or skin absorption. Wear appropriate gloves and safety glasses.

Amphotericin B may be fatal if ingested. It is an irritant and may be harmful by inhalation, ingestion, or skin absorption. Wear appropriate gloves and safety glasses and use in a chemical fume hood. Do not breathe the dust.

Ampicillin may be harmful by inhalation, ingestion, or skin absorption. Wear appropriate gloves and safety glasses and use in a chemical fume hood.

Amplify (fluorographic reagent) contains sodium hydroxide. *See* **Sodium hydroxide.**

Amyl alcohol is extremely flammable and may be harmful by inhalation, ingestion, or skin absorption. It may cause irritation to the skin, eyes, and respiratory tract and affects the central nervous system. Use only with adequate ventilation. Wear appropriate gloves and safety glasses. Keep away from heat, sparks, and open flame.

Anesthetics: Follow manufacturer's safety guidelines

Animal treatment: Procedures for the humane treatment of animals must be observed at all times. Consult your local animal facility for guidelines.

Anisole is an irritant with a foul odor. Avoid contact with the skin or eyes and do not breathe the vapors. Wear appropriate gloves and safety glasses and use in a chemical fume hood.

Antifade, *see* **Phenylenediamine**

Antipain may be harmful by inhalation, ingestion, or skin absorption. Wear appropriate gloves and safety glasses and use in a chemical fume hood.

Aprotinin may be harmful by inhalation, ingestion, or skin absorption. It may also cause allergic reactions. Exposure may cause gastrointestinal effects, muscle pain, blood pressure changes, or bronchospasm. Wear appropriate gloves and safety glasses and use only in a chemical fume hood. Do not breathe the dust.

APTES (Aminopropyltriethoxysilane), *see* **Silane**

Arc lamps are potentially explosive. Follow manufacturer's guidelines. When turning on arc lamps, make sure nearby computers are turned off to avoid damage from electromagnetic wave components. Computers may be restarted once the arc lamps are in operation.

Argon is a nonflammable high-pressure gas. It may be harmful by inhalation, ingestion, or skin absorption. Wear appropriate gloves and safety goggles. Use with sufficient ventilation and do not breathe the gas.

Arsenic is extremely toxic and corrosive and is carcinogenic. It may be fatal by inhalation, ingestion, or skin absorption. Wear appropriate gloves and safety goggles. Do not breathe the vapors.

Arsenic acid is extremely toxic and corrosive and may be carcinogenic. It may be fatal by inhalation, ingestion, or skin absorption. Wear appropriate gloves and safety goggles. Do not breathe the vapors.

Aspartic acid is a possible mutagen and poses a risk of irreversible effects. It may be harmful by inhalation, ingestion, or skin absorption. Wear appropriate gloves and safety glasses. Do not breathe the dust.

ATA, *see* **3-Amino-1,2,4-triazole**

Azahypoxanthine is toxic and may be harmful by inhalation, ingestion, or skin absorption. Avoid contact with the solid or solution. Wear appropriate gloves and safety glasses and use in a chemical fume hood.

Azauracil is toxic and may be harmful by inhalation, ingestion, or skin absorption. Avoid contact with the solid or solution. Wear appropriate gloves and safety glasses and use in a chemical fume hood.

Azide, *see* **Sodium azide**

p-**Azidophenacyl bromide** is highly flammable and corrosive. It may be harmful by inhalation, ingestion, or skin absorption. Wear appropriate gloves and safety goggles and use only in a chemical fume hood. Do not breathe the dust. Keep away from heat, sparks, and open flame.

2,2′-Azino-bis(3-ethylbenzthiazoline)6-sulfonic acid (ABTS) causes irritation to the skin, eyes, and respiratory system. It may be harmful by inhalation, ingestion, or skin absorption. Wear appropriate gloves and safety glasses. Do not breathe the dust.

Bacillus subtilis, *see* **Bacterial strains**

Bacterial strains (shipping of): The Department of Health, Education, and Welfare (HEW) has classified various bacteria into different categories with regard to shipping requirements. Nonpathogenic strains of *E. coli* (such as K12) and *B. subtilis* are in Class 1 and are considered to present no or minimal hazard under normal shipping conditions. However, *Salmonella, Haemophilus,* and certain strains of *Streptomyces* and *Pseudomonas* are in Class 2. Class 2 bacteria are "Agents of ordinary potential hazard: agents which produce disease of varying degrees of severity...but which are contained by ordinary laboratory techniques." See the handbook edited by J.Y. Richmond and R.W. McKinney. 1999. *Biosafety in microbiological and biomedical laboratories (BMBL)*, 4th Edition. U.S. Department of Health and Human Services, Centers for Disease Control at:

http://www.cdc.gov/od/ohs/biosfty/biosfty.htm

BAPTA, 1,2-*bis*(2-aminophenoxy)-ethane-*N,N,N′,N′*-tetraacetic acid, may be harmful by inhalation, ingestion, or skin absorption. It is irritating to mucous membranes, upper respiratory tract, the skin, and the eyes. Wear appropriate gloves and safety glasses and use in a chemical fume hood.

BCIG, *see* **5-Bromo-4-chloro-3-indolyl-β-ᴅ-galactopyranoside**

BCIP, *see* **5-Bromo-4-chloro-3-indolyl-phosphate**

BCP, *see* **1-Bromo-3-chloropropane**

Benzene, C_6H_6, is toxic, a possible carcinogen and mutagen, and flammable. It is volatile and penetrates the skin and is an irritant. It may be harmful by inhalation, ingestion, or skin absorption. Wear appropriate gloves and safety goggles and use in a chemical fume hood. Avoid breathing the vapors.

Benzidine is carcinogenic and may be harmful by inhalation, ingestion, or skin absorption. Avoid breathing the dust. Wear appropriate gloves and safety glasses and handle with care.

Benzoic acid is an irritant and may be harmful by inhalation, ingestion, or skin absorption. Wear appropriate gloves and safety glasses. Do not breathe the dust.

Benzyl alcohol is an irritant and may be harmful by inhalation, ingestion, or skin absorption. Wear appropriate gloves and safety glasses. Keep away from heat, sparks, and open flame.

Benzylaminopurine may be harmful by inhalation, ingestion, or skin absorption. Wear appropriate gloves and safety glasses. It is highly flammable. Keep away from heat, sparks, and open flame.

Benzyl benzoate is an irritant and may be harmful by inhalation, ingestion, or skin absorption. Avoid contact with the eyes. Wear appropriate gloves and safety glasses.

Benzyl viologen may be fatal if inhaled, ingested, or absorbed through the skin. It is irritating to mucous membranes and upper respiratory tract. Wear appropriate gloves and safety glasses and use in a chemical fume hood.

BFA, *see* **BrefeldinA**

Bicinchoninic acid is an irritant. Wear appropriate gloves and goggles and use in a chemical fume hood. Avoid contact with the eyes and skin, and do not breathe the vapors.

Biotin may be harmful by inhalation, ingestion, or skin absorption. Wear appropriate gloves and safety glasses and use in a chemical fume hood.

Biotinylated agarose may be harmful by inhalation, ingestion, or skin absorption. Wear appropriate gloves and safety glasses. Use with adequate ventilation.

Bisacrylamide is a potent neurotoxin and is absorbed through the skin (the effects are cumulative). Avoid breathing the dust. Wear appropriate gloves and a face mask when weighing powdered acrylamide and methylene-bisacrylamide.

Bisbenzimide may be harmful by inhalation, ingestion, or skin absorption. Wear appropriate gloves and safety glasses and use in a chemical fume hood. Do not breathe the dust.

Bleach (Sodium hypochlorite), NaOCl, is poisonous, can be explosive, and may react with organic solvents. It may be fatal by inhalation and is also harmful by ingestion and destructive to the skin. Wear appropriate gloves and safety glasses and use in a chemical fume hood to minimize exposure and odor.

Blood (human) and blood products and Epstein-Barr virus. Human blood, blood products, and tissues may contain occult infectious materials such as hepatitis B virus and HIV that may result in laboratory-acquired infections. Investigators working with EBV-transformed lymphoblast cell lines are also at risk of EBV infection. Any human blood, blood products, or tissues should be considered a biohazard and should be handled accordingly. Wear disposable appropriate gloves, use mechanical pipetting devices, work in a biological safety cabinet, protect against the possibility of aerosol generation, and disinfect all waste materials before disposal. Autoclave contaminated plasticware before disposal; autoclave contaminated liquids or treat with bleach (10% [v/v] final concentration) for at least 30 minutes before disposal. Consult the local institutional safety officer for specific handling and disposal procedures.

BLOTTO (Bovine Lacto Transfer Technique Optimizer) contains sodium azide. *See* **Sodium azide.**

Blue-green lasers present a hazard due to photothermal coagulation. Blue and green wavelengths are readily absorbed by blood hemoglobin.

Bolton-Hunter reagent (radioiodinated). *See also* **Isotope ^{125}I.** Consult the local radiation safety office for further guidance in the appropriate use and disposal of this radioactive material prior to initiating any experiments. Wear appropriate gloves when handling radioactive substances and work only in a radioiodine fume hood. The isotope ^{125}I accumulates in the thyroid and is a potential health hazard.

Boric acid, H$_3$BO$_3$, may be harmful by inhalation, ingestion, or skin absorption. Wear appropriate gloves and goggles.

Bouin's fixative contains formaldehyde, which can cause cancer. It also causes eye and skin burns as well as severe respiratory irritation. It is harmful by inhalation, ingestion, or skin absorption. Wear appropriate gloves and safety goggles and always use in a chemical fume hood.

Bradford dye contains phosphoric acid and methanol. It is corrosive and toxic. Wear appropriate gloves and safety glasses.

BrdU, *see* **5-Bromo-2´-deoxyuridine**

BrefeldinA (BFA) is toxic and highly flammable. It is harmful by inhalation, ingestion, or skin absorption. Wear appropriate gloves and safety goggles and use in a chemical fume hood.

5-Bromo-4-chloro-3-indolyl-β-D-galactopyranoside (BCIG; X-gal) is toxic to the eyes and skin and may be harmful by inhalation, ingestion, or skin absorption. Wear appropriate gloves and safety goggles.

5-Bromo-4-chloro-3-indolyl-β-D-glucuronide (X-Gluc) may be harmful by inhalation, ingestion, or skin absorption. Wear appropriate gloves and safety glasses. Do not breathe the dust.

5-Bromo-4-chloro-3-indolyl-phosphate (BCIP) is toxic and may be harmful by inhalation, ingestion, or skin absorption. Wear appropriate gloves and safety glasses. Do not breathe the dust.

1-Bromo-3-chloropropane (BCP) has a narcotic effect and may be harmful by inhalation, ingestion, or skin absorption. Wear appropriate gloves and safety glasses. Do not breathe the vapor.

5-Bromo-2´-deoxyuridine (BrdU) is a mutagen. It may be harmful by inhalation, ingestion, or skin absorption. It may cause irritation. Avoid breathing the dust. Wear appropriate gloves and safety glasses and always use in a chemical fume hood.

Bromophenol blue may be harmful by inhalation, ingestion, or skin absorption. Wear appropriate gloves and safety glasses and use in a chemical fume hood.

α-**Bungarotoxin** is a potent neurotoxin and is harmful by inhalation, ingestion, or skin absorption. Wear appropriate gloves and safety glasses and always use in a chemical fume hood. Do not breathe the dust.

n-**Butanol** is irritating to the mucous membranes, upper respiratory tract, skin, and especially the eyes. Avoid breathing the vapors. Wear appropriate gloves and safety glasses and use in a chemical fume hood. *n*-Butanol is also highly flammable. Keep away from heat, sparks, and open flame.

sec-**Butanol** is irritating to the mucous membranes, upper respiratory tract, the skin, and especially the eyes. Avoid breathing the vapors. Wear appropriate gloves and safety glasses and use in a chemical fume hood. *sec*-Butanol is also highly flammable. Keep away from heat, sparks, and open flame.

n-**Butyl acetate** is an irritant and may be harmful by inhalation, ingestion, or skin absorption. It poses a risk of serious damage to the eyes. Wear appropriate gloves and safety goggles and use only in a chemical fume hood. Keep away from heat, sparks, and open flame.

Butyl alcohol-*tert* is flammable and may be harmful by inhalation, ingestion, or skin absorption. Wear appropriate gloves and safety glasses. Keep away from heat, sparks, and open flame.

Butyl chloride is highly flammable. Keep away from heat, sparks, and open flame. It may be harmful by inhalation, ingestion, or skin absorption. Wear appropriate gloves and safety goggles. Do not breathe the vapor.

Butyric acid is corrosive and may be harmful by inhalation, ingestion, or skin absorption. Wear appropriate gloves and goggles and use in a chemical fume hood.

Butyrylthiocholine iodide may be harmful by inhalation, ingestion, or skin absorption. Wear appropriate gloves and safety glasses and use in a chemical fume hood.

[14]C, *see* **Radioactive substances**

C$_{18}$ resin, *see* **Resins**

CaC$_2$, *see* **Calcium carbide**

CaCl$_2$, *see* **Calcium chloride**

Cacodylate contains arsenic, is highly toxic, and may be fatal if inhaled, ingested, or absorbed through the skin. Wear appropriate gloves and safety glasses and use in a chemical fume hood. *See also* **Sodium cacodylate**.

Cacodylic acid is toxic and a possible carcinogen. It may cause heritable genetic damage and is harmful by inhalation, ingestion, or skin absorption. Wear appropriate gloves and safety glasses and use only in a chemical fume hood. Do not breathe the dust.

Calcicludine is a poison and may be fatal if it enters the bloodstream. It is harmful by inhalation, ingestion, or skin absorption. Wear appropriate gloves and safety goggles and use in a chemical fume hood. Do not breathe the dust. Do not use if skin is cut or scratched.

Calcium carbide, CaC$_2$, may be harmful by inhalation, ingestion, or skin absorption. Wear appropriate gloves and safety glasses and use in a chemical fume hood. Reaction with water releases acetylene, a flammable gas.

Calcium chloride, CaCl$_2$, is hygroscopic and may cause cardiac disturbances. It may be harmful by inhalation, ingestion, or skin absorption. Do not breathe the dust. Wear appropriate gloves and safety goggles.

Calcium nitrate, Ca(NO$_3$)$_2$, is a strong oxidizer and reacts violently upon contact with many organic substances. Handle with great care. It may be harmful by inhalation, ingestion, or skin absorption. Wear appropriate gloves and safety glasses. Keep away from heat, sparks, and open flame.

Calcium oxide is highly corrosive and causes severe irritation and burns to every area of contact. It may be harmful by inhalation, ingestion, or skin absorption. Wear appropriate gloves and safety goggles.

Calcium sulfate, *see* **Sulfuric acid**

cAMP, *see* **Adenosine 3′,5′-cyclic monophosphate**

Camptothecin is toxic if ingested. It may also be harmful by inhalation or skin absorption. Wear appropriate gloves and safety glasses and use in a chemical fume hood. Do not breathe the dust.

Ca(NO$_3$)$_2$, *see* **Calcium nitrate**

Caproic acid, *see* **Hexanoic acid**

CAPS, *see* **3-(Cyclohexylamino)-1-propanesulfonic acid**

Carbenicillin may cause sensitization by inhalation, ingestion, or skin absorption. Wear appropriate gloves and safety glasses.

Carbon dioxide, CO$_2$, in all forms may be fatal by inhalation, ingestion, or skin absorption. In high concentrations, it can paralyze the respiratory center and cause suffocation. Use only in well-ventilated areas. Contact with carbon dioxide in the form of dry ice can also cause frostbite. Do not place large quantities of dry ice in enclosed areas such as cold rooms. Wear appropriate gloves and safety goggles.

Carbonyl cyanide-*M*-chlorophenyl-hydrazone (CCCP) may be harmful by inhalation or ingestion. Thermal decomposition will produce cyanide (poison). Wear appropriate gloves and use in a chemical fume hood.

1, 1´-Carbonyldiimidazole is extremely destructive to the tissues of the mucous membranes and upper respiratory tract. It may be harmful by inhalation, ingestion, or skin absorption. Wear appropriate gloves and safety glasses.

Carica papaya latex, *see* **Papain**

CBR, *see* **Coomassie brilliant blue**

CCCP, *see* **Carbonyl cyanide-*M*-chlorophenyl-hydrazone**

Celite (diatomaceous earth) contains crystalline silica and poses a lung cancer hazard through prolonged exposure by inhalation. Severe respiratory symptoms may lead to death. It is also an irritant to the eyes and skin. Wear appropriate safety gloves and glasses. Use only with adequate ventilation and do not breathe the dust.

Cellulase is an irritant and may cause allergic reactions. It may be harmful by inhalation, ingestion, or skin absorption. Wear appropriate gloves and safety glasses. Do not breathe the dust.

Cesium chloride, CsCl, may be harmful by inhalation, ingestion, or skin absorption. Wear appropriate gloves and safety glasses.

Cesium hydroxide, CsOH, may be harmful by inhalation, ingestion, or skin absorption. It is extremely destructive to the mucous membranes and upper respiratory tract; inhalation may be fatal. Do not breathe the dust. Wear appropriate gloves and safety glasses and always use in a chemical fume hood.

Cesium iodide, CsI, poses a possible risk to the unborn child and causes severe irritation to the thyroid. It may be harmful by inhalation, ingestion, or skin absorption. Do not breathe the dust. Wear appropriate gloves and safety glasses and always use in a chemical fume hood.

Cetylpyridinium bromide (CPB) causes severe irritation to the eyes, skin, and respiratory tract. Wear appropriate gloves and safety glasses and use in a chemical fume hood.

Cetyltrimethylammonium bromide (CTAB) is toxic and an irritant and may be harmful by inhalation, ingestion, or skin absorption. Wear appropriate gloves and safety glasses. Avoid breathing the dust.

C$_6$H$_6$, *see* **Benzene**

CHAPS, *see* **3-[(3-Cholamidopropyl)dimethyl-ammonio]-1-propanesulfonate**

C$_6$H$_5$CH$_3$, *see* **Toluene**

CHCl$_3$, *see* **Chloroform**

C$_2$H$_4$Cl$_2$, *see* **1,2-Dichloroethane**

CH$_2$Cl$_2$, *see* **Dichloromethane**

CH$_3$CH$_2$OH, *see* **Ethanol**

C$_2$H$_4$INO, *see* **Iodoacetamide**

Chicago Sky Blue is a possible mutagen and may be harmful by inhalation, ingestion, or skin absorption. Wear appropriate gloves and safety glasses and use in a chemical fume hood. Do not breathe the dust.

C$_7$H$_7$FO$_2$S, *see* **Phenylmethylsulfonyl fluoride**

(CH$_3$)$_2$SO$_4$, *see* **Dimethyl sulfate**

(C$_2$H$_5$)$_2$SO$_4$, *see* **Diethyl sulfate**

Chloral hydrate is extremely destructive to tissues of the mucous membranes and upper respiratory tract. It may be harmful by inhalation, ingestion, or skin absorption. Wear appropriate gloves and safety glasses and use in a chemical fume hood.

Chloramine is corrosive and causes burns. It may be harmful by inhalation, ingestion, or skin absorption. Wear appropriate gloves and safety glasses and use in a chemical fume hood. Do not breathe the dust.

Chloramphenicol is a potential carcinogen and may be harmful by inhalation, ingestion, or skin absorption. Wear appropriate gloves and safety glasses and use in a chemical fume hood.

Chloroform, CHCl$_3$, is irritating to the skin, eyes, mucous membranes, and respiratory tract. It is a carcinogen and may damage the liver and kidneys. It is also volatile. Avoid breathing the vapors. Wear appropriate gloves and safety glasses and always use in a chemical fume hood.

p-**Chloromercuriphenylsulfonic acid** may be fatal by inhalation, ingestion, or skin absorption. Wear appropriate gloves and safety goggles and use in a chemical fume hood.

4-Chloro-1-naphthol is irritating to the eyes, skin, mucous membranes, and respiratory tract. Handle with care. Wear appropriate gloves and safety glasses.

Chloroquine may be harmful by inhalation, ingestion, or skin absorption. Prolonged exposure can lead to permanent eye damage. Wear appropriate gloves and safety goggles.

N-**Chlorosuccinimide** may be harmful by inhalation, ingestion, or skin absorption. Wear appropriate gloves and safety glasses and use in a chemical fume hood.

3-[(3-Cholamidopropyl)dimethyl-ammonio]-1-propanesulfonate (CHAPS) is an irritant and may be harmful by inhalation, ingestion, or skin absorption. Wear appropriate gloves and safety glasses.

Cholesteryl chloroformate may be harmful by inhalation, ingestion, or skin absorption. Wear appropriate gloves and safety glasses and use in a chemical fume hood.

Chromic acid is corrosive and extremely dangerous by inhalation, ingestion, or skin absorption. Wear appropriate gloves and safety goggles. Keep away from heat, sparks, and open flame.

Chromogenic substrates may be carcinogenic. Consult your local institutional safety officer for handling and disposal procedures.

Chymostatin is an irritant and may be harmful by inhalation, ingestion, or skin absorption. Do not breathe the dust.

Citraconic anhydride is highly toxic and may be harmful by inhalation, ingestion, or skin absorption. Wear appropriate gloves and safety goggles. Avoid inhalation.

Citric acid is an irritant and may be harmful by inhalation, ingestion, or skin absorption. It poses a risk of serious damage to the eyes. Wear appropriate gloves and safety goggles. Do not breathe the dust.

CO_2, *see* **Carbon dioxide**

Cobalt chloride, $CoCl_2$, may be harmful by inhalation, ingestion, or skin absorption. Wear appropriate gloves and safety glasses.

Cobalt nitrate, $Co(NO_3)_2 \cdot 6H_2O$, is a strong oxidizer and may be harmful by inhalation, ingestion, or skin absorption. Wear appropriate gloves and safety goggles.

Cocaine is a controlled substance available to researchers who have a valid DEA license and an approved protocol. It is highly toxic and affects the central nervous system. It is harmful by inhalation, ingestion, or skin absorption. Wear appropriate gloves and safety glasses. Do not breathe the dust.

CoCl$_2$, *see* **Cobalt chloride**

Colcemid is highly toxic and may cause organ failure or death if inhaled or swallowed. It may cause reproductive or fetal effects. It is harmful by inhalation, ingestion, or skin absorption. Wear appropriate gloves and safety goggles and use only in a chemical fume hood. Do not breathe the dust.

Colchicine is highly toxic, may be fatal, and may cause cancer and heritable genetic damage. It may be harmful by inhalation, ingestion, or skin absorption. Wear appropriate gloves and safety goggles and use only in a chemical fume hood. Do not breathe the dust. Do not use if skin is cut or scratched.

Concanavalin A, biotinylated, may be harmful by inhalation, ingestion, or skin absorption. Wear appropriate gloves and safety goggles and use in a chemical fume hood.

ω**-Conotoxin** is a poison and may be fatal if it enters the bloodstream. It may be harmful by inhalation, ingestion, or skin absorption. Wear appropriate gloves and safety goggles and use in a chemical fume hood. Do not breathe the dust. Do not use if skin is cut or scratched.

Co(NO$_3$)$_2$•6H$_2$O, *see* **Cobalt nitrate**

Coomassie brilliant blue (CBR) may be harmful by inhalation, ingestion, or skin absorption. Wear appropriate gloves and safety glasses.

Copper chloride, CuCl$_2$, is toxic and an irritant. It may be harmful by inhalation, ingestion, or skin absorption. Wear appropriate gloves and safety glasses and use in a chemical fume hood. Do not breathe the dust.

Copper sulfate, CuSO$_4$, may be harmful by inhalation or ingestion. Wear appropriate gloves and safety glasses.

*p***-Coumaric acid** is an irritant and may be harmful by inhalation, ingestion, or skin absorption. Wear appropriate gloves and safety glasses.

CPB, *see* **Cetylpyridinium bromide**

m-**Cresol** may be fatal if inhaled, ingested, or absorbed through the skin. It is corrosive and is extremely destructive to the eyes, skin, mucous membranes, and upper respiratory tract. Wear appropriate gloves and safety glasses and use in a chemical fume hood.

p-**Cresol** may be fatal if inhaled, ingested, or absorbed through the skin. It is corrosive and is extremely destructive to the eyes, skin, mucous membranes, and upper respiratory tract. Wear appropriate gloves and safety glasses and use in a chemical fume hood.

Crystal violet can cause severe burns. It may be harmful by inhalation, ingestion, or skin absorption. Wear appropriate gloves and safety goggles and use in a chemical fume hood. Do not breathe the dust.

CsCl, *see* **Cesium chloride**

CsI, *see* **Cesium iodide**

CsOH, *see* **Cesium hydroxide**

CTAB, *see* **Cetyltrimethylammonium bromide**

CuCl$_2$, *see* **Copper chloride**

Curare is highly toxic and may be fatal if swallowed. It targets the nerves and skeletal muscles. It is harmful by inhalation, ingestion, or skin absorption. Wear appropriate gloves and safety glasses. Do not breathe the dust.

CuSO$_4$, *see* **Copper sulfate**

Cyanoacrylate adhesives are harmful by inhalation, ingestion, and skin absorption. Immediate bonding of tissues can occur. Do not pull apart. Inhalation may cause lightheadedness. Wear appropriate gloves and safety glasses. Follow manufacturer's safety guidelines.

Cyanogen bromide is extremely toxic and volatile. It may be fatal by inhalation, ingestion, or skin absorption. Do not breathe the vapors. Wear appropriate gloves and always use in a chemical fume hood. Keep away from acids.

α-**Cyano-4-hydroxycinnamic acid** may cause cardiac disturbances. Chronic effects may be delayed. It may be harmful by inhalation, ingestion, or skin absorption. Wear appropriate gloves and safety glasses.

β-Cyanoethyl diisopropylchlorophosphoramidite, 2-Cyanoethyl diiso-propylchlorophosphoramidite, is highly corrosive and causes burns. Heating may cause an explosion. It is extremely destructive to mucous membranes and upper respiratory tract, eyes, and skin. Wear appropriate gloves and safety goggles and use in a chemical fume hood. Do not breathe the vapor. Keep away from heat, sparks, and open flame.

Cycloheximide may be fatal if inhaled, ingested, or absorbed through the skin. Wear appropriate gloves and safety glasses and use in a chemical fume hood.

3-(Cyclohexylamino)-1-propanesulfonic acid (CAPS) is an irritant and may be harmful by inhalation, ingestion, or skin absorption. Wear appropriate gloves and safety glasses. Do not breathe the dust.

Cyclophosphamide is a mutagen and a carcinogen. It may be fatal if inhaled, ingested, or absorbed through the skin. It may also cause irritation. Avoid breathing the vapors. Wear appropriate gloves and safety glasses and use in a chemical fume hood. The ISOPAC vial is designed to limit exposure to this substance. Consult the local institutional safety officer for appropriate procedures for discarding the ISOPAC vial and other contaminated materials.

Cyclosporin is a powerful immunosuppressant and possible carcinogen. Avoid chronic exposure. It may be harmful by inhalation, ingestion, or skin absorption. Wear appropriate gloves and safety glasses. Use only in a chemical fume hood. Do not breathe the dust.

Cysteine is an irritant to the eyes, skin, and respiratory tract. It may be harmful by inhalation, ingestion, or skin absorption. Wear appropriate gloves and safety glasses. Do not breathe the dust.

Cysteine hydrochloride is an irritant to the eyes, skin, and respiratory tract. It may be harmful by inhalation, ingestion, or skin absorption. Wear appropriate gloves and safety glasses. Do not breathe the dust.

Cytochalasin B may be fatal by inhalation, ingestion, or skin absorption, and is a possible teratogen with possible irreversible effects. Do not breathe the dust. Wear appropriate gloves and safety goggles and use only in a chemical fume hood.

DAB, *see* **3,3´-Diaminobenzidine**

DABCO, *see* **1,4-Diazabicyclo-[2,2,2]-octane**

DAPI, *see* **4,6-Diamidine-2-phenylindole dihydrochloride**

DCM, *see* **Dichloromethane**

DDAB, *see* **Didecyldimethylammonium bromide**

DEAE, *see* **Diethylaminoethanol**

DEDTC, *see* **Diethyldithiocarbamate**

Demecolcine, *see* **Colchicine; Colcemid**

Deoxycholic acid sodium salt (DOC) may be harmful by inhalation, ingestion, or skin absorption. Do not breathe the dust. Wear appropriate gloves and safety glasses.

DEPC, *see* **Diethyl pyrocarbonate**

DES, *see* **Diethyl sulfate**

Deuterated compounds can affect biochemical processes. Ingestion of large amounts can impair kidney function and central nervous system operation.

Developing jelly is caustic. Wear appropriate gloves. Do not get on skin or clothes. If jelly gets on skin, wash immediately and thoroughly with soap and water.

Dexamethasone may be harmful by inhalation, ingestion, or skin absorption. Overexposure may cause reproductive disorder(s) and possible risks of harm to the unborn child. Wear appropriate gloves and safety glasses. Do not breathe the dust.

DFP, *see* **Diisopropyl fluorophosphate**

DHB, *see* **2,5-Dihydroxybenzoic acid**

4,6-Diamidine-2 phenylindole dihydrochloride (DAPI) is a possible carcinogen. It may be harmful by inhalation, ingestion, or skin absorption. It may also cause irritation. Avoid breathing the dust and vapors. Wear appropriate gloves and safety glasses and use in a chemical fume hood.

3,3′-Diaminobenzidine (DAB) is a carcinogen. Handle with extreme care. Avoid breathing vapors. Wear appropriate gloves and safety glasses and use in a chemical fume hood.

1,6-Diaminohexane may be harmful by inhalation, ingestion, or skin absorption. Wear appropriate gloves and safety glasses.

1,3-Diamino-2-hydroxypropanol-*N,N,N′,N′*-tetraacetic acid (DPTA) is an irritant and may be harmful by inhalation, ingestion, or skin absorption. Wear appropriate gloves and safety glasses. Do not breathe the dust.

1,3-Diaminopropane may be fatal if absorbed through the skin. It is highly toxic and corrosive. The liquid and vapor are both flammable. It is harmful by inhalation, ingestion, or skin absorption. Avoid breathing the dust and vapors. Wear appropriate gloves and safety glasses and use in a chemical fume hood. Keep away from heat, sparks, and open flame.

1,4-Diazabicyclo-[2,2,2]-octane (DABCO) may be harmful by inhalation, ingestion, or skin absorption. Wear appropriate gloves and safety glasses and use in a chemical fume hood.

Diazomethane is extremely flammable and may be harmful by inhalation, ingestion, or skin absorption. Wear appropriate gloves and safety glasses. Do not breathe the vapors.

Dibutyl phthalate may be harmful by inhalation, ingestion, or skin absorption. Wear appropriate gloves and safety glasses. Do not breathe the vapors.

Dichlorodimethylsilane is extremely flammable. The vapor may cause flash fire. It may be fatal by inhalation, ingestion, or skin absorption and causes severe burns to every area of contact. Wear appropriate gloves and safety goggles and always use in a chemical fume hood.

1,2-Dichloroethane, $C_2H_4Cl_2$, may be harmful by inhalation, ingestion, or skin absorption and is a carcinogen. Wear appropriate gloves and safety glasses and use in a chemical fume hood. *Flammable:* Vapor may travel considerable distances to source of ignition and flash back. Keep away from heat, sparks, and open flame.

Dichloromethane (DCM), CH$_2$Cl$_2$, (also known as **Methylene chloride)** is toxic if inhaled, ingested, or absorbed through the skin. It is also an irritant and is suspected to be a carcinogen. Wear appropriate gloves and safety glasses and use in a chemical fume hood. Do not breathe the vapors.

Dichloromethylsilane, *see* **Dichlorodimethylsilane; Dichlorosilane**

2,4-Dichlorophenoxyacetic acid (2,4-D) may cause liver and kidney damage. It may be harmful by inhalation, ingestion, or skin absorption. Wear appropriate gloves and safety glasses.

Dichlorosilane is highly flammable and toxic and may be fatal if inhaled. It is harmful by inhalation, ingestion, or skin absorption. Wear appropriate gloves and safety goggles and use only in a chemical fume hood. It reacts violently with water. Keep away from heat, sparks, and open flame. Take precautionary measures against static discharges.

Dichromic acid, H$_2$Cr$_2$O$_7$, contains chromium (VI), a cancer hazard. It is a strong oxidizer and may be harmful by inhalation or ingestion. It causes irritation and may cause rash or external ulcers and may be fatal through skin absorption. Wear appropriate gloves, goggles, and protective clothing. Use with adequate ventilation and wash thoroughly after handling.

Didecyldimethylammonium bromide (DDAB) is toxic and may be harmful by inhalation, ingestion, or skin absorption. Wear appropriate gloves and safety glasses.

DIEA, *see* **Diisopropylethylamine**

Diethanolamine may be harmful by inhalation, ingestion, or skin absorption. Wear appropriate gloves and safety glasses.

Diethylamine, NH(C$_2$H$_5$)$_2$, is corrosive, toxic, and extremely flammable. It may be harmful by inhalation, ingestion, or skin absorption. Wear appropriate gloves and safety glasses and use only in a chemical fume hood. Keep away from heat, sparks, and open flame.

Diethylaminoethanol (DEAE) may be harmful by inhalation, ingestion, or skin absorption. Wear appropriate gloves and safety glasses and use in a chemical fume hood.

Diethyldithiocarbamate (DEDTC) may be harmful by inhalation, ingestion, or skin absorption. Wear appropriate gloves and safety glasses.

Diethyl ether, Et$_2$O or (C$_2$H$_5$)$_2$O, is extremely volatile and flammable. It is irritating to the eyes, mucous membranes, and skin. It is also a CNS depressant with anesthetic effects. It may be harmful by inhalation, ingestion, or skin absorption. Avoid breathing the vapors. Wear appropriate gloves and safety glasses and always use in a chemical fume hood. Explosive peroxides can form during storage or upon exposure to air or direct sunlight. Keep away from heat, sparks, and open flame.

Diethylene ether, *see* **1,4-Dioxane**

Diethyl pyrocarbonate (DEPC) is a potent protein denaturant and is a suspected carcinogen. Aim bottle away from you when opening it; internal pressure can lead to splattering. Wear appropriate gloves, safety goggles, and lab coat and use in a chemical fume hood.

Diethyl sulfate (DES), (C$_2$H$_5$)$_2$SO$_4$, is a mutagen and a probable carcinogen. It is also volatile. Avoid breathing the vapors. Wear appropriate gloves and use in a chemical fume hood. Use screw-cap tubes for all DES-treated cultures, and mechanical pipettors to manipulate DES solutions. Dispose of all DES-treated cultures in bleach. For disposal procedures of DES solutions, *see* **Dimethyl sulfate**.

Digitonin may be fatal if inhaled, ingested, or absorbed through the skin. Wear appropriate gloves and safety glasses and use in a chemical fume hood.

Digoxigenin may be fatal if inhaled, ingested, or absorbed through the skin. Wear appropriate gloves and safety glasses and use in a chemical fume hood.

2,5-Dihydroxybenzoic acid (DHB) may be harmful by inhalation, ingestion, or skin absorption. Wear appropriate gloves and safety glasses. Do not breathe the dust.

Diisopropylcarbodiimide (DIPC) may be fatal if inhaled, ingested, or absorbed through the skin. It is highly toxic and irritating to the mucous membranes, the upper respiratory tract, skin, and especially the eyes. It may cause allergic respiratory and skin reactions. Wear appropriate gloves and safety glasses and always use in a chemical fume hood. Keep away from heat, sparks, and open flame.

Diisopropylethylamine (DIEA) (DIPEA) is extremely destructive to the mucous membranes, upper respiratory tract, skin, and eyes. It may be harmful by ingestion or skin absorption. Inhalation may be fatal. Wear appropriate gloves and safety glasses and always use in a chemical fume hood. Keep away from heat, sparks, and open flame.

Diisopropyl fluorophosphate (DFP) is an extremely toxic cholinesterase inhibitor. It is much more effective than phenylmethylsulfonyl fluoride (PMSF), but due to its toxicity, DFP should not be used in the laboratory. If it must be used, avoid any contact with the skin and wear full protective clothing, appropriate gloves, and safety goggles. Use only in a chemical fume hood.

***N,N*-Dimethylacetamide (DMA)** is toxic and may cause harm to the unborn child. It may be harmful by inhalation, ingestion, or skin absorption. Wear appropriate gloves and safety glasses and use in a chemical fume hood. Do not breathe the mist or vapor. It is also flammable. Keep away from heat, sparks, and open flame.

4-Dimethylaminopyridine (DMAP) is highly toxic and corrosive. It may be harmful by inhalation, ingestion, or skin absorption. Wear appropriate gloves and safety glasses. Handle with great care.

Dimethyl benzyl ammonium chloride may be harmful by inhalation, ingestion, or skin absorption. Wear appropriate gloves and safety glasses and use in a chemical fume hood.

Dimethyldichlorosilane is extremely flammable and corrosive, therefore causing severe burns. It may be harmful by inhalation, ingestion, or skin absorption. Wear appropriate gloves and safety goggles and use only in a well-ventilated area.

***N,N*-Dimethylethylenediamine** may be harmful by inhalation, ingestion, or skin absorption. Wear appropriate gloves and safety glasses and use in a chemical fume hood.

N,N-**Dimethylformamide (DMF), HCON(CH₃)₂**, is a possible carcinogen and is irritating to the eyes, skin, and mucous membranes. It can exert its toxic effects through inhalation, ingestion, or skin absorption. Chronic inhalation can cause liver and kidney damage. Wear appropriate gloves and safety glasses and use in a chemical fume hood.

4,5-Dimethoxy-2-nitrobenzaldehyde (DMNB) is a possible cancer-causing agent and may be harmful by inhalation, ingestion, or skin absorption. Wear appropriate gloves and safety glasses and use in a chemical fume hood.

Dimethyl pimelimidate (DMP) is irritating to the eyes, skin, mucous membranes, and upper respiratory tract. It can exert harmful effects by inhalation, ingestion, or skin absorption. Avoid breathing the vapors. Wear appropriate gloves, face mask, and safety glasses.

Dimethyl sulfate (DMS), (CH₃)₂SO₄, is extremely toxic and is a carcinogen. Avoid breathing the vapors. Wear appropriate gloves and safety glasses and use only in a chemical fume hood. Dispose of solutions containing dimethyl sulfate by pouring them slowly into a solution of sodium hydroxide or ammonium hydroxide and allowing them to sit overnight in the chemical fume hood. Contact your local safety office before reentering the lab to clean up a spill.

Dimethyl sulfide (DMS), *see* **Dimethyl sulfate**

Dimethyl sulfoxide (DMSO) may be harmful by inhalation or skin absorption. Wear appropriate gloves and safety glasses and use in a chemical fume hood. DMSO is also combustible. Store in a tightly closed container. Keep away from heat, sparks, and open flame.

2,4-Dinitrofluorobenzene (DNFB) is a highly toxic agent suspected of being a carcinogen. It may be harmful by inhalation, ingestion, or skin absorption. Wear appropriate gloves and safety glasses. Handle with great care.

Dinitrophenol (DNP) is toxic and highly flammable. It may be fatal by inhalation, ingestion, or skin absorption. Wear appropriate gloves and safety glasses and use only in a chemical fume hood. Keep away from heat, sparks, and open flame.

Dioctylphthalate is toxic and is suspected of being a carcinogen. It may be harmful by inhalation, ingestion, or skin absorption. Wear appropriate gloves and safety goggles.

1,4-Dioxane is highly flammable both in liquid and vapor form. It is a possible carcinogen and is highly toxic by inhalation, ingestion, or skin absorption. Do not breathe the vapor. Wear appropriate gloves and safety glasses. Keep away from heat, sparks, and open flame.

DIPEA, *see* **Diisopropylethylamine**

DIPC, *see* **1,3-Diisopropylcarbodiimide**

Diphenyloxazole (PPO) may be carcinogenic. It may be harmful by inhalation, ingestion, or skin absorption. Wear appropriate gloves and safety glasses. Consult your local institutional safety officer for specific handling and disposal procedures.

Diphtheria toxin is a biohazard and may be fatal if it enters the bloodstream. It may be harmful by inhalation, ingestion, or skin absorption. Wear appropriate gloves and safety goggles.

2,2´-Dipyridyl disulfide, *see* **2,2´-Dithiopyridine**

Disodium citrate, *see* **Citric acid**

5,5´-Dithiobis-(2-nitrobenzoic acid) (DTNB) is an irritant and may be harmful by inhalation, ingestion, or skin absorption. Wear appropriate gloves and safety glasses. Do not breathe the dust.

2,2´-Dithiopyridine (2,2´-Dipyridyl disulfide) may be harmful by inhalation, ingestion, or skin absorption. Wear appropriate gloves and safety glasses.

Dithiothreitol (DTT) is a strong reducing agent that emits a foul odor. It may be harmful by inhalation, ingestion, or skin absorption. When working with the solid form or highly concentrated stocks, wear appropriate gloves and safety glasses and use in a chemical fume hood.

DMA, *see* **N,N-Dimethylacetamide**

DMAP, *see* **4-Dimethylaminopyridine**

DMF, *see* **N,N-Dimethylformamide**

DMNB, *see* **4,5-Dimethoxy-2-nitrobenzaldehyde**

DMP, *see* **Dimethyl pimelimidate**

DMS, *see* **Dimethyl sulfate; Dimethyl sulfide**

DMSO, *see* **Dimethyl sulfoxide**

DNFB, *see* **2,4-Dinitrofluorobenzene**

DNP, *see* **Dinitrophenol**

DOC, *see* **Deoxycholic acid sodium salt**

Dopamine is an irritant and may be harmful by inhalation, ingestion, or skin absorption. Wear appropriate gloves and safety glasses. Do not breathe the dust.

DPTA, *see* **1,3-Diamino-2-hydroxypropanol-*N,N,N´,N´*,-tetraacetic acid**

DPX is composed of Distyrene, a plasticizer, and xylene, and is commercially available. Follow the manufacturer's guidelines for handling DPX.

Drierite is toxic, may be fatal, and is dangerous to the environment. It may cause cancer by inhalation. It is harmful by inhalation, ingestion, or skin absorption. Wear appropriate gloves and safety goggles. Do not breathe the dust.

Dry ice, *see* **Carbon dioxide**

DTNB, *see* **5,5´-Dithiobis-(2-nitrobenzoic acid)**

DTT, *see* **Dithiothreitol**

Dyes: Follow manufacturer's safety guidelines

EDC, *see* ***N*-ethyl-*N´*-(dimethylaminopropyl)-carbodiimide**

EDCI (Ethyl-3-[3-dimethylamino] propyl carbodiimide), *see* ***N*-Ethyl-*N´*-(dimethylaminopropyl)-carbodiimide**

EDPA, *see* **Ethylenediamine-N,N´-diacetic acid**

Ellman's reagent, *see* **5,5´-Dithiobis-(2-nitrobenzoic acid) (DTNB)**

EMS, *see* **Ethyl methane sulfonate**

Entellan may be fatal if ingested and can cause harm to many organs. It is highly flammable and may be harmful by inhalation, ingestion, or skin absorption. Wear appropriate gloves and safety goggles and use in a chemical fume hood. Keep away from heat, sparks, and open flame.

ENU, *see N*-**Nitroso-*N*-ethylurea (ENU)**

Epichlorohydrin (1-chloro-2,3-epoxypropane) can cause burns to any area of contact. It is also highly flammable in both liquid and vapor forms. Keep away from heat, sparks, and open flame. It is a possible carcinogen and a possible birth defect hazard. It may be harmful by inhalation, ingestion, or skin absorption. Wear appropriate gloves and safety goggles and use in a chemical fume hood.

Epinephrine may be fatal if inhaled, ingested, or absorbed through the skin. Wear appropriate gloves and safety glasses.

Epon resin, *see* **Resins**

Epoxy and acrylic resins, *see* **Resins**

Escherichia coli (E. coli), *see* **Bacterial strains**

Ethanedithiol is toxic and harmful by inhalation, ingestion, or skin absorption. Handle with care and avoid any contact with skin. Wear appropriate gloves and goggles and use only in a chemical fume hood.

Ethanol (EtOH), CH_3CH_2OH, is highly flammable and may be harmful by inhalation, ingestion, or skin absorption. Wear appropriate gloves and safety glasses. Keep away from heat, sparks, and open flame.

Ethanolamine, $HOCH_2CH_2NH_2$, is toxic and harmful by inhalation, ingestion, or skin absorption. Handle with care and avoid any contact with the skin. Wear appropriate gloves and goggles and use in a chemical fume hood. Ethanolamine is highly corrosive and reacts violently with acids.

Ether, *see* **Diethyl ether**

Ethidium bromide is a powerful mutagen and is toxic. Consult the local institutional safety officer for specific handling and disposal procedures. Avoid breathing the dust. Wear appropriate gloves when working with solutions that contain this dye.

N-(Ethoxycarbonylmethyl)-6-methoxyquinolinium bromide (MQAE) is potentially harmful by inhalation, ingestion, or skin absorption. Wear appropriate gloves and safety glasses.

Ethyl acetate may be fatal by ingestion and harmful by inhalation or skin absorption. Wear appropriate gloves and safety goggles. Do not breathe the dust. Use in a well-ventilated area.

1-Ethyl-3-[3-dimethylaminopropyl] carbodiimide (EDC), *see N*-Ethyl-*N′*-(dimethylaminopropyl)-carbodiimide

N-Ethyl-*N′*-(dimethylaminopropyl)-carbodiimide (EDC) is irritating to the mucous membranes and upper respiratory tract. It may be harmful by inhalation, ingestion, or skin absorption. Wear appropriate gloves and safety glasses. Handle with care.

Ethyl-3-[3-dimethylamino] propyl carbodiimide (EDCI), *see N*-Ethyl-*N′*-(dimethylaminopropyl)-carbodiimide

Ethylene is extremely cold and may cause burns similar to frostbite. It may be harmful by inhalation (which may lead to chemical asphyxiation), ingestion, or skin absorption. Wear appropriate gloves and safety glasses. Keep away from heat, sparks, and open flame.

Ethylenediamine-N,N′-diacetic acid (EDPA) is an irritant and may be harmful by inhalation, ingestion, or skin absorption. Wear appropriate gloves and safety glasses.

Ethylene glycol may be harmful by inhalation, ingestion, or skin absorption. Wear appropriate gloves and safety glasses and use in a chemical fume hood.

N-Ethylmaleimide (NEM) may be harmful by inhalation, ingestion, or skin absorption. Wear appropriate gloves and safety glasses and always use in a chemical fume hood.

Ethyl methane sulfonate (EMS) is a volatile organic solvent that is a mutagen and carcinogen. It is harmful if inhaled, ingested, or absorbed through the skin. Discard supernatants and washes containing EMS in a beaker containing 50% sodium thiosulfate. Decontaminate all material that has come in contact with EMS by treatment in a large volume of 10% (w/v) sodium thiosulfate. Use extreme caution when handling. When using undiluted EMS, wear protective appropriate gloves and use in a chemical fume hood. Store EMS in the cold. DO NOT mouth-pipette EMS. Pipettes used with undiluted EMS should not be too warm; chill them in the refrigerator before use to minimize the volatility of EMS. All glassware coming in contact with EMS should be immersed in a large beaker of 1 N NaOH or laboratory bleach before recycling or disposal.

N-Ethylmorpholine is highly flammable in liquid or vapor forms. It may be harmful by inhalation, ingestion, or skin absorption. Wear appropriate gloves and safety goggles. Keep away from heat, sparks, and open flame.

Ethylnitrosourea (ENU), *see* **N-Nitroso-N-ethylurea**

EtOH, *see* **Ethanol**

Et₂O or (C₂H₅)₂O, *see* **Diethyl ether**

Euparol is highly flammable and will be easily ignited by heat, sparks, or open flame. It may be harmful by inhalation, ingestion, or skin absorption. The vapors may cause dizziness or suffocation. Wear appropriate gloves and safety goggles and work in a well-ventilated area.

Exonuclease III may be harmful by inhalation, ingestion, or skin absorption. Wear appropriate gloves and safety goggles.

Fast Green is a carcinogen and may be harmful by inhalation, ingestion, or skin absorption. Wear appropriate gloves and safety glasses and always use in a chemical fume hood.

Fast Red may cause methemoglobinemia through overexposure. It may be harmful by inhalation, ingestion, or skin absorption. Wear appropriate gloves and safety glasses.

FeCl₃, *see* **Ferric chloride**

Ferric chloride, FeCl₃, may be harmful by inhalation, ingestion, or skin absorption. Wear appropriate gloves and safety glasses and use only in a chemical fume hood.

Fialuridine (FIAU) may be extremely harmful by inhalation, ingestion, or skin absorption. Wear appropriate gloves and safety glasses and use only in a chemical fume hood.

FIAU, *see* **Fialuridine**

FITC, *see* **Fluorescein isothiocyanate**

Fixatives: Follow manufacturer's safety guidelines

Flash lamps, due to their light intensity, can be harmful to the eyes. They also may explode on occasion. Wear appropriate eye protection and follow manufacturer's guidelines.

Fluorescein isothiocyanate (FITC) may be harmful by inhalation, ingestion, or skin absorption. Wear appropriate gloves and safety glasses.

Fluorescein sodium salt may be harmful by inhalation, ingestion, or skin absorption. Wear appropriate gloves and safety glasses and use in a chemical fume hood.

Fluoric acid is extremely hazardous in both liquid and vapor forms. It is corrosive and poisonous and may be fatal. It is harmful by inhalation, ingestion, or skin absorption. Wear appropriate gloves and safety goggles and use in a chemical fume hood. Do not breathe the vapor. Keep away from heat, sparks, and open flame.

Fluoromount-G contains sodium azide, which is very toxic if ingested or inhaled. It is highly poisonous and blocks the cytochrome electron transport system. Solutions containing sodium azide should be clearly marked. Wear appropriate gloves and safety goggles and handle sodium azide with great care.

Fluoropyrimidine compounds are toxic and may be harmful by inhalation, ingestion, or skin absorption. Avoid direct contact. Wear appropriate gloves and safety glasses and use only in a chemical fume hood.

Formaldehyde, HCHO, is highly toxic and volatile. It is also a possible carcinogen. It is readily absorbed through the skin and is irritating or destructive to the skin, eyes, mucous membranes, and upper respiratory tract. Avoid breathing the vapors. Wear appropriate gloves and safety glasses and always use in a chemical fume hood. Keep away from heat, sparks, and open flame.

Formalin is a solution of formaldehyde in water. *See* **Formaldehyde**

Formamide is teratogenic. The vapor is irritating to the eyes, skin, mucous membranes, and upper respiratory tract. It may be harmful by inhalation, ingestion, or skin absorption. Wear appropriate gloves and safety glasses and always use a chemical fume hood when working with concentrated solutions of formamide. Keep working solutions covered as much as possible.

Formic acid, HCOOH, is highly toxic and extremely destructive to tissue of the mucous membranes, upper respiratory tract, eyes, and skin. It may be harmful by inhalation, ingestion, or skin absorption. Wear appropriate gloves and safety glasses (or face shield) and use in a chemical fume hood.

Freund's adjuvants are potentially harmful. Be careful during preparation and injection.

G418 (an aminoglycosidic antibiotic) is toxic and may cause harm to the unborn child. It may be harmful by inhalation, ingestion, or skin absorption. Wear appropriate gloves and safety goggles and use in a chemical fume hood. Do not breathe the dust.

β**-Galactosidase** is an irritant and may cause allergic reactions. It may be harmful by inhalation, ingestion, or skin absorption. Wear appropriate gloves and safety glasses.

Gamborg's B5 powder medium may be toxic and cause systemic poisoning. The effects may occur very rapidly. Contact with combustible material may cause fire. It may be harmful by inhalation, ingestion, or skin absorption. Wear appropriate gloves and safety glasses. Do not breathe the dust. Contact lenses should not be worn.

Gamma rays, *see* **Radioactive substances**

Ganciclovir (GCV) is highly toxic, may cause heritable genetic damage, and may impair fertility. It may be harmful by inhalation, ingestion, or skin absorption. Wear appropriate gloves and safety goggles. Do not breathe the dust.

GCV, *see* **Ganciclovir**

Giemsa may be fatal or cause blindness by ingestion and is toxic by inhalation and skin absorption. There is a possible risk of irreversible effects. Wear appropriate gloves and safety goggles and use only in a chemical fume hood. Do not breathe the dust.

Glacial acetic acid, *see* **Acetic acid (glacial)**

Glassware, pressurized, must be used with extreme caution. Autoclave and cool sealed bottles in metal containers, pressurize bottles behind Plexiglas shields, and encase 20-liter bottles in wire mesh. Handle glassware under vacuum, such as desiccators, vacuum traps, drying equipment, or a reactor for working under argon atmosphere, with appropriate caution. Always wear safety glasses.

Glass wool may be harmful by inhalation and may cause skin irritation. Wear appropriate gloves and mask.

Gluconic acid may be harmful by inhalation, ingestion, or skin absorption. Wear appropriate gloves and safety glasses.

Glucose oxidase may be harmful by inhalation, ingestion, or skin absorption. Wear appropriate gloves and safety glasses.

β-Glucuronidase (GUS) may be harmful by inhalation, ingestion, or skin absorption. Wear respirator, appropriate gloves, and safety glasses.

Glufosinate ammonium may be harmful by inhalation, ingestion, or skin absorption. Wear respirator, appropriate gloves, and safety glasses. Do not breathe the dust.

Glutaraldehyde is toxic. It is readily absorbed through the skin and is irritating or destructive to the skin, eyes, mucous membranes, and upper respiratory tract. Wear appropriate gloves and safety glasses and always use in a chemical fume hood.

Glycine may be harmful by inhalation, ingestion, or skin absorption. Wear gloves and safety glasses. Avoid breathing the dust.

Gold chloride solution is an irritant and may be harmful by inhalation, ingestion, or skin absorption. Wear gloves and safety glasses. Avoid breathing the dust.

Gonadotropin is a possible teratogen and poses a risk of irreversible effects. It may be harmful by inhalation, ingestion, or skin absorption. Wear appropriate gloves and safety goggles and use in a chemical fume hood. Do not breathe the dust.

Guanidine hydrochloride is irritating to the mucous membranes, upper respiratory tract, skin, and eyes. It may be harmful by inhalation, ingestion, or skin absorption. Wear appropriate gloves and safety glasses. Avoid breathing the dust.

Guanidine thiocyanate may be harmful by inhalation, ingestion, or skin absorption. Wear appropriate gloves and safety glasses.

Guanidinium hydrochloride, *see* **Guanidine hydrochloride**

Guanidinium isothiocyanate, *see* **Guanidine thiocyanate**

Guanidinium thiocyanate, *see* **Guanidine thiocyanate**

Gum arabic causes severe eye irritation and may cause allergic or respiratory reaction. It may form combustible dust concentrations in air. It may be harmful by inhalation, ingestion, or skin absorption. Wear appropriate gloves and safety goggles. Do not breathe the dust.

GUS, *see* **β-Glucuronidase**

^{3}H, *see* **Radioactive substances**

H_2, *see* **Hydrogen**

Halothane is an anesthetic and may be harmful by inhalation or skin absorption. It is also highly volatile. Wear appropriate gloves and safety glasses and work in a well-ventilated area.

HAME, *see p*-**Hydroxybenzoic acid methyl ester**

H_3BO_3, *see* **Boric acid**

HBTU, *see* **2-[1H-benzotriazol-1-yl]-1,1,3,3-tetramethyluronium hexafluorophosphate**

2-[1H-benzotriazol-1-yl]-1,1,3,3-tetramethyluronium hexafluorophosphate (HBTU) is irritating to the eyes, skin, mucous membranes, and upper respiratory tract. It may be harmful by inhalation, ingestion, or skin absorption. Wear appropriate gloves and safety glasses and do not breathe the dust.

H₃CCOONH₄, *see* **Ammonium acetate**

HCl, *see* **Hydrochloric acid**

HCHO, *see* **Formaldehyde**

H₃COH, *see* **Methanol**

HCON(CH₃)₂, *see* **Dimethylformamide**

HCOOH, *see* **Formic acid**

H₂Cr₂O₇, *see* **Dichromic acid**

Helium (gas) may cause frostbite and may be harmful by inhalation, ingestion, or skin absorption. Wear appropriate gloves and safety goggles. Do not wear contact lenses.

Hematoxylin, Harris Modified, may be harmful by inhalation, ingestion, or skin absorption. Wear appropriate gloves and safety goggles.

Hematoxylin, Mayer's, may be harmful by inhalation, ingestion, or skin absorption. Wear appropriate gloves and safety goggles.

Heptafluorobutyric acid (HFBA) is corrosive and may be harmful by inhalation, ingestion, or skin absorption. Wear appropriate gloves and safety goggles.

Heptane may be harmful by inhalation, ingestion, or skin absorption. Wear appropriate gloves and safety glasses. It is extremely flammable. Keep away from heat, sparks, and open flame.

1-Hexadecene is an irritant and may be harmful by inhalation, ingestion, or skin absorption. Wear appropriate gloves and safety glasses. Do not breathe the vapors. Keep away from heat, sparks, and open flame.

Hexane is extremely flammable and may be harmful by inhalation, ingestion, or skin absorption. Wear appropriate gloves and safety glasses and use
only in a chemical fume hood. Keep away from heat, sparks, and open flame.

1,6-Hexanediamine is corrosive and may be harmful by inhalation, ingestion, or skin absorption. Wear appropriate gloves and safety goggles.

Hexanoic acid (ε-Amino-*n*-caproic acid; Caproic acid) is extremely
destructive to tissues of the mucous membranes and upper respiratory tract,
eyes, and skin. It may be harmful by inhalation, ingestion, or skin absorption. Wear appropriate gloves and safety glasses.

HFBA, *see* **Heptafluorobutyric acid**

Hg, *see* **Mercury**

Historesin, *see* **Xylene**

HNO$_3$, *see* **Nitric acid**

H$_2$NOH, *see* **Hydroxylamine**

H$_2$O$_2$, *see* **Hydrogen peroxide**

HOBT, *see* **1-Hydroxybenzotriazole**

HOCH$_2$CH$_2$NH$_2$, *see* **Ethanolamine**

HOCH$_2$CH$_2$SH, *see* **β-Mercaptoethanol**

Hoechst No. 33258, *see* **Bisbenzimide**

Hoechst No. 33342, *see* **Bisbenzimide**

H$_3$PO$_2$, *see* **Hypophosphorous acid**

H$_3$PO$_4$, *see* **Phosphoric acid**

H$_2$S, *see* **Hydrogen sulfide**

H$_2$SO$_4$, *see* **Sulfuric acid**

Hydrazine, N₂H₄, is highly toxic and explosive in the anhydrous state. It may be harmful by inhalation, ingestion, or skin absorption. Avoid breathing the vapors. Wear appropriate gloves, goggles, and protective clothing, and use only in a chemical fume hood. Dispose of solutions containing hydrazine in accordance with MSDS recommendations. Keep away from heat, sparks, and open flame.

Hydrobromic acid is toxic, corrosive, and causes severe burns. It may be harmful by inhalation, ingestion, or skin absorption. Wear appropriate gloves and safety goggles and use only in a chemical fume hood. Do not breathe the vapor.

Hydrochloric acid, HCl, is volatile and may be fatal if inhaled, ingested, or absorbed through the skin. It is extremely destructive to mucous membranes, upper respiratory tract, eyes, and skin. Wear appropriate gloves and safety glasses and use with great care in a chemical fume hood. Wear goggles when handling large quantities.

Hydrofluoric acid is extremely toxic, corrosive, and can cause severe burns. It may be harmful by inhalation, ingestion, and skin absorption. Wear appropriate gloves and safety goggles and use only in a chemical fume hood.

Hydrogen, H₂, is an odorless, highly flammable gas that is lighter than air and, at concentrations above 4%, forms explosive mixtures with air. Handle it with respect and caution. Keep away from heat, sparks, and open flame.

Hydrogen fluoride is extremely toxic, corrosive, and can cause severe burns. It may be harmful by inhalation, ingestion, and skin absorption. Wear appropriate gloves and safety goggles and use only in a chemical fume hood.

Hydrogen peroxide, H₂O₂, is corrosive, toxic, and extremely damaging to the skin. It may be harmful by inhalation, ingestion, or skin absorption. Wear appropriate gloves and safety glasses and use only in a chemical fume hood.

Hydrogen sulfide, H₂S, is an extremely toxic gas that causes paralysis of the respiratory center. It is irritating and corrosive to tissues and may cause olfactory fatigue. Do not rely on odor to detect its presence. Take great care when handling it. Keep H₂S tanks in a chemical fume hood or a room equipped with appropriate ventilation. Wear appropriate gloves and safety glasses. It is also very flammable. Keep away from heat, sparks, and open flame.

Hydrophosphorous acid is corrosive and extremely destructive to the mucous membranes, respiratory tract, eyes, and skin. It is harmful by inhalation, ingestion, or skin absorption. Wear appropriate gloves and safety goggles and use in a chemical fume hood.

Hydroquinone, *see* **Benzene**

Hydroxybenzoic acid is an irritant and may be harmful by inhalation, ingestion, or skin absorption. Wear appropriate gloves and safety glasses. Do not breathe the dust.

p-**Hydroxybenzoic acid methyl ester (HAME)** is an irritant and may be harmful by inhalation, ingestion, or skin absorption. Wear appropriate gloves and safety glasses. Do not breathe the dust.

1-Hydroxybenzotriazole (HOBT) and **N-Hydroxybenzotriazole** are highly flammable and irritating to the eyes, skin, and mucous membranes. They may be toxic by inhalation, ingestion, or skin absorption. Wear appropriate gloves and safety glasses and use in a chemical fume hood. Keep away from heat, sparks, and open flame.

Hydroxylamine, H_2NOH, is corrosive and toxic. It may be harmful by inhalation, ingestion, or skin absorption. Wear appropriate gloves and safety glasses and use only in a chemical fume hood.

Hydroxylamine hydrochloride, H_2NOH HCl, is corrosive and extremely toxic. This substance is a blood toxin and a possible mutagen. It may be harmful by inhalation, ingestion, or skin absorption. Wear appropriate gloves and safety goggles and use only in a chemical fume hood. Do not breathe the dust.

2-Hydroxyquinine is an irritant and may be harmful by inhalation, ingestion, or skin absorption. Wear appropriate gloves and safety glasses.

Hydroxyquinoline is irritating to the eyes, skin, mucous membranes, and upper respiratory tract. It may be harmful by inhalation, ingestion, or skin absorption. Do not breathe the dust. Wear appropriate gloves and safety glasses.

N-**Hydroxysuccinimide** is an irritant and may be harmful by inhalation, ingestion, or skin absorption. Wear appropriate gloves and safety glasses.

Hydroxyurea is toxic and may cause heritable genetic damage and harm to the unborn child. It may be harmful by inhalation, ingestion, or skin absorption. Wear appropriate gloves and safety glasses and use only in a chemical fume hood. Do not breathe the dust.

Hygromycin B is highly toxic and may be fatal if inhaled, ingested, or absorbed through the skin. Wear appropriate gloves and safety goggles and use only in a chemical fume hood. Do not breathe the dust.

Hypophosphorous acid, H_3PO_2, is usually supplied as a 50% solution, which is corrosive and should be handled with care. It should be freshly diluted immediately before use. Wear appropriate gloves and safety glasses and use in a chemical fume hood.

[125]I, *see* **Radioactive substances**

[131]I, *see* **Radioactive substances**

IAA, *see* **Isoamyl alcohol**

ICR 191 must be handled like any potentially dangerous mutagen. Avoid breathing the vapors. Wear protective appropriate gloves. ICR 191 should be weighed out in the same manner as MNNG. First, weigh a glass scintillation vial or its equivalent (~16 g) on a Mettler balance and record its weight. Working under a hood, transfer a small amount of ICR 191 to the vial with a spatula. Reweigh the vial and determine the weight of the ICR 191. All tubes and glassware coming in contact with ICR 191 should be disposed of or washed in accordance with MSDS recommendations.

Imidazole is corrosive and may be harmful by inhalation, ingestion, or skin absorption. Wear appropriate gloves and safety glasses and use in a chemical fume hood.

Iminodiacetic acid is an irritant and may be harmful by inhalation, ingestion, or skin absorption. Wear appropriate gloves and safety goggles. Do not breathe the dust.

Inositol may be harmful by inhalation, ingestion, or skin absorption. Wear appropriate gloves and safety glasses.

Iodine crystals are volatile at room temperature and form a purple vapor that is intensely irritating to the eyes and mucous membranes. Always wear appropriate gloves and safety glasses when handling iodine, and use in a chemical fume hood.

Iodoacetamide, C_2H_4INO, can alkylate amino groups in proteins and can therefore cause problems if the antigen is being purified for amino acid sequencing. It is toxic and harmful by inhalation, ingestion, or skin absorption. Wear appropriate gloves and safety glasses and use only in a chemical fume hood. Do not breathe the dust.

Iodoacetic acid, *see* **Acetic acid**

Iodogen may be harmful by inhalation, ingestion, or skin absorption. Wear appropriate gloves and safety goggles. Do not breathe the dust.

o-**Iodosobenzoic acid** is an irritant and may be harmful by inhalation, ingestion, or skin absorption. Wear appropriate gloves and safety goggles.

2-IP solution is an irritant and may be harmful by inhalation, ingestion, or skin absorption. Wear appropriate gloves and safety glasses.

IPTG, *see* **Isopropyl-β-D-thiogalactopyranoside**

Isoamyl acetate is flammable and an irritant. It may be harmful by inhalation, ingestion, or skin absorption. Wear appropriate gloves and safety goggles. Do not breathe the vapors or mist. Keep away from heat, sparks, and open flame.

Isoamyl alcohol (IAA) may be harmful by inhalation, ingestion, or skin absorption, and presents a risk of serious damage to the eyes. Wear appropriate gloves and safety goggles. Keep away from heat, sparks, and open flame.

Isobutanol, *see* **Isobutyl alcohol**

Isobutyl alcohol (Isobutanol) is extremely flammable and may be harmful by inhalation or ingestion. Wear appropriate gloves and safety glasses. Keep away from heat, sparks, and open flame.

Isofluorane (Isoflurane) is an irritant and may be harmful by inhalation, ingestion, or skin absorption. Chronic exposure may be harmful. Wear appropriate gloves and safety glasses.

Isopentane (2-Methylbutane) is extremely flammable. Keep away from heat, sparks, and open flame. It may be harmful by inhalation, ingestion, or skin absorption. Wear appropriate gloves and safety glasses.

Isopropanol is flammable and irritating. It may be harmful by inhalation, ingestion, or skin absorption. Wear appropriate gloves and safety glasses. Do not breathe the vapor. Keep away from heat, sparks, and open flame.

Isopropyl-β-D-thiogalactopyranoside (IPTG) may be harmful by inhalation, ingestion, or skin absorption. Wear appropriate gloves and safety glasses.

Isopsoralen, *see* **Psoralen**

Isotope ^{125}I accumulates in the thyroid and is a potential health hazard. Consult the local radiation safety office for further guidance in the appropriate use and disposal of radioactive materials. Wear appropriate gloves when handling radioactive substances. The ^{125}I$_2$ formed during oxidation of Na^{125}I is volatile. Work in an approved chemical fume hood with a charcoal filter when exposing the Na^{125}I to oxidizing reagents such as chloramine-T, IODO-GEN, or acids. Because the oxidation proceeds very rapidly and releases large amounts of volatile ^{125}I$_2$ when chloramine-T is used, it is important to be well prepared for each step of the reaction, so that the danger of contamination from volatile radiation can be minimized. Shield all forms of the isotope with lead. When handling the isotope, wear one or two pairs of appropriate gloves, depending on the amount of isotope being used and the difficulty of the manipulation required.

Kanamycin may be harmful by inhalation, ingestion, or skin absorption. Wear appropriate gloves and safety glasses. Use only in a well-ventilated area.

KBr, *see* **Potassium bromide**

KCl, *see* **Potassium chloride**

KCN, *see* **Potassium cyanide**

Kerosene may be fatal if swallowed and is flammable in both liquid and vapor forms. It may be harmful by inhalation, ingestion, or skin absorption.

Wear appropriate gloves and safety goggles. Use only in a well-ventilated area. Keep away from heat, sparks, and open flame.

Ketoprofen is toxic. It is harmful by inhalation, ingestion, or skin absorption. Wear appropriate gloves and safety goggles and always use in a chemical fume hood.

$K_3Fe(CN)_6$, *see* **Potassium ferricyanide**

$K_4Fe(CN)_6$•3H_2O, *see* **Potassium ferrocyanide**

K-Gluconate, *see* **Potassium gluconate**

KI, *see* **Potassium iodide**

$KMnO_4$, *see* **Potassium permanganate**

KNO_3, *see* **Potassium nitrate**

KOH, *see* **Potassium hydroxide**

KSCN, *see* **Potassium thiocyanate**

Lactic acid is corrosive and causes severe irritation and burns to any area of contact. It may be harmful by inhalation, ingestion, or skin absorption. Wear appropriate gloves and safety goggles. Do not breathe vapor or mist.

Laser radiation, both visible and invisible, can be seriously harmful to the eyes and skin and may generate airborne contaminants, depending on the class of laser used. High-power lasers cause permanent eye damage and can burn exposed skin, ignite flammable materials, and activate toxic chemicals that release hazardous by-products. Avoid eye or skin exposure to direct or scattered radiation. Do not stare at the laser and do not point the laser at anyone. Wear appropriate eye protection and use suitable shields that are designed to offer protection for the specific type of wavelength, mode of operation (continuous wave or pulsed), and power output (watts) of the laser being used. Avoid wearing jewelry or other objects that may reflect or scatter the beam. Some non-beam hazards include electrocution, fire, and asphyxiation. Entry to the area in which the laser is being used must be controlled and posted with warning signs that indicate when the laser is in use. Always follow suggested safety guidelines that accompany the equipment and contact your local safety office for further information.

Ion lasers present a hazard due to high-voltage, high-current power supplies. Always follow manufacturer's suggested safety guidelines.

Ultraviolet lasers present a hazard due to invisible beam, high-energy radiation. Always use beam traps, scattered light shields, and fluorescent beamfinder cards.

Blue-green lasers present a hazard due to photothermal coagulation. Blue and green wavelengths are readily absorbed by blood hemoglobin.

L-DOPA is irritating to the eyes, skin, and respiratory system. It may be harmful by inhalation, ingestion, or skin absorption. Wear appropriate gloves and safety glasses.

N-**Lauroylsarcosine** is an irritant and may be harmful by inhalation, ingestion, or skin absorption. Wear appropriate gloves and safety glasses.

Lead, Pb, is a toxic metal. It presents a long-term danger, because lead accumulates in the liver and interferes with its function. Avoid contact with skin and wear appropriate gloves when handling.

Lead citrate is a potential cancer hazard and may cause possible death. There is a danger of cumulative effects. It is harmful by inhalation, ingestion, or skin absorption. Wear appropriate gloves and safety goggles and always use in a chemical fume hood. Do not breathe the dust.

Lead nitrate, $Pb(NO_3)_2$, and **all other lead salts** may be harmful by inhalation, ingestion, or skin absorption. Wear appropriate gloves and safety glasses and always use in a chemical fume hood.

Leupeptin (or its **hemisulfate**) may be harmful by inhalation, ingestion, or skin absorption. Wear appropriate gloves and safety glasses and use in a chemical fume hood.

LiCl, *see* **Lithium chloride**

Liquid nitrogen (LN_2) can cause severe damage due to extreme temperature. Handle frozen samples with extreme caution. Do not breathe the vapors. Seepage of liquid nitrogen into frozen vials can result in an exploding tube upon removal from liquid nitrogen. Use vials with O-rings when possible. Wear cryo-mitts and a face mask. No not allow the liquid nitrogen to spill onto your clothes. Do not breathe the vapors.

Lithium acetate may be harmful by inhalation, ingestion, or skin absorption. Wear appropriate gloves and safety glasses and use in a chemical fume hood. Do not breathe the dust.

Lithium chloride, LiCl, is an irritant to the eyes, skin, mucous membranes, and upper respiratory tract. It may be harmful by inhalation, ingestion, or skin absorption. Wear appropriate gloves, safety goggles, and use in a chemical fume hood. Do not breathe the dust.

Lithium dodecyl sulfate may be harmful by ingestion, inhalation, or skin absorption. Wear appropriate gloves when weighing, and use in a chemical fume hood.

LN$_2$, *see* **Liquid nitrogen**

Lowicryl is an irritant and may be harmful by inhalation, ingestion, or skin absorption. Wear appropriate gloves and safety glasses.

LR white resin, *see* **Resins**

Luminol is an irritant and may be harmful by inhalation, ingestion, or skin absorption. Wear appropriate gloves and safety glasses.

Lysozyme is caustic to mucous membranes. Wear appropriate gloves and safety glasses.

Magnesium chloride, MgCl$_2$, may be harmful by inhalation, ingestion, or skin absorption. Wear appropriate gloves and safety glasses and use in a chemical fume hood.

Magnesium sulfate, MgSO$_4$, may be harmful by inhalation, ingestion, or skin absorption. Wear appropriate gloves and safety glasses and use in a chemical fume hood.

Malachite green may be harmful by inhalation, ingestion, or skin absorption. Wear appropriate gloves and safety glasses and use in a chemical fume hood.

Maleic acid is toxic and harmful by inhalation, ingestion, or skin absorption. Reaction with water or moist air can release toxic, corrosive, or flammable gases. Do not breathe the vapors or dust. Wear appropriate gloves and safety glasses.

Maleimide is extremely harmful and may be fatal by inhalation, ingestion, or skin absorption. Do not breathe the dust. Wear appropriate gloves and safety goggles and use in a chemical fume hood.

***N*-[gamma-Maleimidobutyryloxy]sulfo succinimide ester,** *see* **Sulfo-GMBS**

***m*-Maleimidobenzoyl sulfo succinimide-ester,** *see* **Sulfo-MBS**

Manganese acetate may be harmful by inhalation, ingestion, or skin absorption. The effects may be delayed. Men exposed to the dusts showed a decrease in fertility. Wear appropriate gloves and safety glasses and use in a chemical fume hood. Do not breathe the dust. Keep away from heat, sparks, and open flame.

Manganese chloride, $MnCl_2$, may be harmful by inhalation, ingestion, or skin absorption. Wear appropriate gloves and safety glasses and use in a chemical fume hood.

Manganese sulfate may be harmful by inhalation, ingestion, or skin absorption. Wear appropriate gloves and safety glasses.

Maxam-Gilbert sequencing: Many of the reagents used in Maxam-Gilbert sequencing are toxic and/or volatile. Carry out all degradation reactions in a well-ventilated chemical fume hood.

MeOH or H_3COH, *see* **Methanol**

β-Mercaptoethanol (2-Mercaptoethanol), $HOCH_2CH_2SH$, may be fatal if inhaled or absorbed through the skin and is harmful if ingested. High concentrations are extremely destructive to the mucous membranes, upper respiratory tract, skin, and eyes. β-Mercaptoethanol has a very foul odor. Wear appropriate gloves and safety glasses and always use in a chemical fume hood.

2-Mercaptoethylamine may be harmful by inhalation, ingestion, or skin absorption. Wear appropriate gloves and safety glasses.

Mercury, Hg, may be fatal if inhaled, ingested, or absorbed through the skin. It presents a long-term danger, since mercury accumulates in the liver and interferes with its function. Wear appropriate gloves and safety glasses and use in a chemical fume hood. Because of its high vapor pressure, spills of mercury should be cleaned immediately using mercury-absorbing reagents.

Merthiolate, *see* **Thimerosal**

MES, *see* **2-(*N*-Morpholino)ethanesulfonic acid**

Methacrylate is toxic and may be harmful by inhalation, ingestion, or skin absorption. Wear appropriate gloves and safety glasses. Do not breathe the vapors.

Methanol, MeOH or **H$_3$COH,** is toxic and can cause blindness. It may be harmful by inhalation, ingestion, or skin absorption. Adequate ventilation is necessary to limit exposure to vapors. Avoid inhaling these vapors. Wear appropriate gloves and safety goggles and use only in a chemical fume hood.

Methanolic-HCl is toxic, can cause blindness, and is highly flammable. It may be harmful by inhalation, ingestion, or skin absorption. Adequate ventilation is necessary to limit exposure to vapors. Avoid inhaling these vapors. Wear appropriate gloves and safety goggles and use only in a chemical fume hood.

Methotrexate (MTX) is a carcinogen and a teratogen. It may be harmful by inhalation, ingestion, or skin absorption. Exposure may cause gastrointestinal effects, bone marrow suppression, or liver or kidney damage. It may also cause irritation. Avoid breathing the vapors. Wear appropriate gloves and safety glasses and always use in a chemical fume hood.

***N*-6-Methoxyquinolyl acetoethyl ester** is a derivative of *N*-(Ethoxy-carbonylmethyl)-6-methoxyquinolinium bromide (MQAE)

Methyl acetate is extremely flammable in both liquid and vapor form. Vapor may cause flash fire. It may be harmful by inhalation, ingestion, or skin absorption. Wear appropriate gloves and safety glasses. Keep away from heat, sparks, and open flame.

2-Methylbutane, *see* **Isopentane**

Methyl cyanide, *see* **Acetonitrile**

Methylformamide is highly toxic, may impair fertility, and may cause harm to the unborn child. It may be harmful by inhalation, ingestion, or skin absorption. Wear appropriate gloves and safety goggles. Do not breathe the vapor.

N,N´-**Methylenebisacrylamide** is a poison and may affect the central nervous system. It may be harmful by inhalation, ingestion, or skin absorption. Wear appropriate gloves and safety glasses. Do not breathe the dust.

Methylene blue is irritating to the eyes and skin. It may be harmful by inhalation, ingestion, or skin absorption. Wear appropriate gloves and safety glasses.

Methylene chloride, *see* **Dichloromethane**

Methyl green may be harmful by inhalation, ingestion, or skin absorption. Wear appropriate gloves and safety glasses and use in a chemical fume hood.

Methylformamide, *see* **Dimethylformamide**

Methyl 4-hydroxybenzoate (Nipagin) is an irritant and may be harmful by inhalation, ingestion, or skin absorption. Wear appropriate gloves and safety glasses.

1-Methylimidazole is highly corrosive and causes burns. It is harmful by inhalation, ingestion, or skin absorption. Wear appropriate gloves and safety goggles and use only in a chemical fume hood.

Methyl iodide may be fatal and can affect the central nervous system. It is also a carcinogen and an irritant. It is harmful by inhalation, ingestion, or skin absorption. Wear appropriate gloves and safety goggles and use only in a chemical fume hood.

Methylmercuric hydroxide is extremely toxic and may be harmful by inhalation, ingestion, or skin absorption. It is also volatile. Therefore, carry out all manipulations of solutions containing concentrations of methylmercuric hydroxide in excess of 10^{-2} M in a chemical fume hood and wear appropriate gloves when handling such solutions. Treat all solid and liquid wastes as toxic materials and dispose of in accordance with MSDS recommendations. Organomercury compounds easily penetrate gloves and skin.

N-**Methyl-***N´*-**nitro-***N*-**nitrosoguanidine (MNNG)** is a mutagen and carcinogen. It may be harmful by inhalation, ingestion, or skin absorption.

Consult your local safety office for specific handling and disposal procedures. Use extreme caution and avoid breathing the vapors. Perhaps the most dangerous part of handling MNNG is the point at which the bottle of MNNG is opened. When the vapor pressure is released, the solid MNNG crystals and powder can be dispersed and inhaled if precautionary measures are not taken. Therefore, always open a bottle of solid MNNG with protective appropriate gloves under a chemical fume hood. One recommended method of weighing out MNNG to prepare a stock solution is to use sterile glass scintillation vials with plastic screw caps, or glass vials used for slants. First, weigh the vial on a Mettler or similar precision balance; typical vials are 16 g or less. Then, under the hood, place a small amount of MNNG in the vial and cover and weigh again. From the added weight, calculate the weight of MNNG. In this manner, the bottle of MNNG is never opened away from the hood and the solid MNNG is always in a covered container. Do not weigh on paper in a Mettler! To prepare a stock solution, ideally about 35–50 mg of MNNG should be placed in a vial. All materials coming in contact with MNNG should be immersed in a large beaker of 1 N NaOH or laboratory bleach prior to recycling or disposal.

1-Methylpiperazine is highly flammable, combustible, and causes burns. It may be harmful by inhalation, ingestion, or skin absorption. Wear appropriate gloves and safety glasses. Do not breathe the dust. Keep away from heat, sparks, and open flame.

(R)-(-)-2-Methylpiperazine is highly flammable and causes burns. It may be harmful by inhalation, ingestion, or skin absorption. Wear appropriate gloves and safety glasses. Do not breathe the dust. Keep away from heat, sparks, and open flame

2-Methylpyridine, *see* α-**Picoline**

N-**Methyl-2-pyrrolidinone, 2-Methyl-2-pyrrolidinone,** and **1-Methyl-2-pyrrolidinone** are combustible and may be harmful by inhalation, ingestion, or skin absorption. Wear appropriate gloves and safety glasses and use in a chemical fume hood. Do not breathe the vapors. Keep away from heat, sparks, and open flame.

Methyl salicylate is volatile and may be harmful by inhalation, ingestion, or skin absorption. Do not breathe the dust. Wear appropriate gloves and safety glasses and use in a chemical fume hood.

4-Methylumbelliferone (4-MU) is an irritant and may be harmful by inhalation, ingestion, or skin absorption. Wear appropriate gloves and safety glasses.

Methyl paraben (*p*-hydroxymethylbenzoate), *see* **Methyl 4-hydroxybenzoate (Nipagin)**

Methyl viologen may be fatal if inhaled, ingested, or absorbed through the skin. It is irritating to mucous membranes and upper respiratory tract. Wear appropriate gloves and safety glasses and use in a chemical fume hood.

Metofane is toxic and may be harmful by inhalation, ingestion, or skin absorption. Wear appropriate gloves and safety glasses.

Metronidazole is an irritant and may be harmful by inhalation, ingestion, or skin absorption. Avoid breathing the vapors as well as direct contact. Wear appropriate gloves and safety glasses.

Mevinolin (Lovastatin) may be irritating to the respiratory tract, eyes, and skin. Wear appropriate gloves and safety glasses and use in a chemical fume hood.

MgCl$_2$, *see* **Magnesium chloride**

MgSO$_4$, *see* **Magnesium sulfate**

Microtome blades are extremely sharp! Use care when sectioning. If unfamiliar with the use of a microtome, have someone demonstrate its use.

Mifepristone is toxic and may cause harm to the unborn child. It may impair fertility and damage the female reproductive system. Wear appropriate gloves and safety goggles. Do not breathe the dust.

Mitomycin C is a carcinogen. It may be fatal by inhalation, ingestion, or skin absorption. Do not breathe the dust. Wear appropriate gloves and safety goggles and use only in a chemical fume hood.

Monensin may be fatal by inhalation, ingestion, or skin absorption. Wear appropriate gloves and safety goggles and use only in a chemical fume hood.

2-[*N*-Morpholino]ethanesulfonic acid (MES) may be harmful by inhalation, ingestion, or skin absorption. Wear appropriate gloves and safety glasses.

3-(*N*-Morpholino)-propanesulfonic acid (MOPS) may be harmful by inhalation, ingestion, or skin absorption. It is irritating to mucous membranes and upper respiratory tract. Wear appropriate gloves and safety glasses and use in a chemical fume hood.

MnCl$_2$, *see* **Manganese chloride**

MNNG, *see* ***N*-Methyl-*N′*-nitro-*N*-nitrosoguanidine**

MOPS, *see* **3-(*N*-Morpholino)-propanesulfonic acid**

MQAE, *see* ***N*-(Ethoxycarbonylmethyl)-6-methoxyquinolinium bromide**

MTX, *see* **Methotrexate**

myo-Insitol-(2-3H) is highly flammable and may be harmful by inhalation, ingestion, or skin absorption. Wear appropriate gloves and safety glasses. Keep away from heat, sparks, and open flame.

NaBH$_4$, *see* **Sodium borohydride**

NaClO$_4$, *see* **Sodium perchlorate**

Na$_2$CO$_3$, *see* **Sodium carbonate**

NaF, *see* **Sodium fluoride**

NaI, *see* **Sodium iodide**

NaIO$_4$, *see* **Sodium periodate**

Nalidixic acid may be harmful by inhalation, ingestion, or skin absorption. Wear appropriate gloves and safety goggles.

NaN$_3$, *see* **Sodium azide**

NaNO$_2$, *see* **Sodium nitrite**

NaNO$_3$, *see* **Sodium nitrate**

NaOAc, *see* **Sodium acetate**

NaOCl, *see* **Bleach**

NaOD, see **Sodium deuteroxide**

NaOH, *see* **Sodium hydroxide**

Naphthyl phosphate may be harmful by inhalation, ingestion, or skin absorption. Wear appropriate gloves and safety glasses.

Na$_2$SO$_4$, *see* **Sodium sulfate**

Na$_2$S$_2$O$_5$, *see* **Sodium metabisulfite**

Na$_3$VO$_4$, *see* **Sodium orthovanadate**

NBT, *see* **4-Nitro blue tetrazolium chloride**

NEM, *see* **N-Ethylmaleimide**

Neomycin may be harmful by inhalation, ingestion, or skin absorption. Wear appropriate gloves and safety glasses.

Neutral red may be harmful by inhalation, ingestion, or skin absorption. Wear appropriate gloves and safety glasses.

NH$_3$, *see* **Ammonia**

N$_2$H$_4$, *see* **Hydrazine**

NH(C$_2$H$_5$)$_2$, *see* **Diethylamine**

NH$_4$Cl, *see* **Ammonium chloride**

(NH$_4$)$_2$CO$_3$, *see* **Ammonium carbonate**

NH$_4$HCO$_3$, *see* **Ammonium bicarbonate**

(NH$_4$)$_6$Mo$_7$O$_{24}$•4H$_2$O, *see* **Ammonium molybdate**

NH$_4$NO$_3$, *see* **Ammonium nitrate**

NH$_4$Ac, *see* **Ammonium acetate**

NH$_4$OH, *see* **Ammonium hydroxide**

(NH$_4$)$_2$SO$_4$, *see* **Ammonium sulfate**

(NH$_4$)$_2$S$_2$O$_8$, *see* **Ammonium persulfate**

Nickel chloride, NiCl$_2$, is toxic and may be harmful by inhalation, ingestion, or skin absorption. Do not breathe the dust. Wear appropriate gloves and safety glasses.

Nickel sulfate, NiSO$_4$, is a possible carcinogen and may cause heritable genetic damage. It is a skin irritant and may be harmful by inhalation, ingestion, or skin absorption. Wear appropriate gloves and safety glasses and use in a chemical fume hood. Do not breathe the dust.

NiCl$_2$, *see* **Nickel chloride**

Nigericin is highly toxic and may be fatal by inhalation, ingestion, or skin absorption. It is irritating to the eyes, skin, mucous membranes, and upper respiratory tract. Wear appropriate gloves and safety goggles and use in a chemical fume hood. Do not breathe the dust.

Ninhydrin may be harmful by inhalation, ingestion, or skin absorption. It is irritating to the eyes, skin, mucous membranes, and upper respiratory tract. Wear appropriate gloves and safety glasses and use in a chemical fume hood.

Ni-NTA resin, *see* **Resins**

Nipagin, *see* **Methyl 4-hydroxybenzoate**

NiSO$_4$, *see* **Nickel sulfate**

Nitric acid, HNO$_3$, is volatile and must be handled with great care. It is toxic by inhalation, ingestion, or skin absorption. Wear appropriate gloves and safety goggles and use in a chemical fume hood. Do not breathe the vapors. Keep away from heat, sparks, and open flame.

4-Nitro blue tetrazolium chloride (NBT) may be harmful by inhalation, ingestion, or skin absorption. Wear appropriate gloves and safety glasses.

Nitrogen (gaseous or liquid) may be harmful by inhalation, ingestion, or skin absorption. Wear appropriate gloves and safety glasses. Consult your local safety office for proper precautions.

p-**Nitrophenyl phosphate (PNPP)** is toxic and may be harmful by inhalation, ingestion, or skin absorption. Wear appropriate gloves and safety glasses and use only in a chemical fume hood.

N-**Nitroso-*N*-ethylurea (ENU)** is a possible carcinogen and is a potent mutagen. It may be harmful by inhalation, ingestion, or skin absorption. Wear appropriate gloves and safety glasses and use in a chemical fume hood. Do not breathe the dust. Decontaminate all material that has been in contact with ENU in 1 N NaOH.

6-Nitroveratryl chloroformate (NVOC-Cl) is corrosive and causes burns to areas of contact. It may be harmful by inhalation, ingestion, or skin absorption. Do not breathe the dust. Wear appropriate gloves and safety goggles and use in a chemical fume hood.

Nocodazole is a possible mutagen. It may be harmful by inhalation, ingestion, or skin absorption. Wear appropriate gloves and safety goggles.

Nonidet P40 causes severe eye irritation and burns. It may be harmful by inhalation, ingestion, or skin absorption. Wear appropriate gloves and safety goggles.

NVOC-Cl, *see* **6-Nitroveratryl chloroformate**

O₂, *see* **Oxygen**

OCT is composed of polyvinyl alcohol, polyethylene glycol, and dimethyl benzyl ammonium chloride. Follow the manufacturer's guidelines for handling OCT.

Octane is highly flammable in both liquid and vapor forms. Keep away from heat, sparks, and open flame. It may be harmful by inhalation, ingestion, or skin absorption. Wear appropriate gloves and safety glasses.

Octanol is highly flammable. Keep away from heat, sparks, and open flame. It may be harmful by inhalation, ingestion, or skin absorption. Wear appropriate gloves and safety goggles.

Octopamine may be harmful by inhalation, ingestion, or skin absorption. Exposure may cause an increase in blood pressure. Wear appropriate gloves and safety glasses. Do not breathe the dust.

OPFP, *see* **Pentafluorophenol** and **Pentafluorophenyl esters**

Orthophosphoric acid, *see* **Phosphoric acid**

Osmium is toxic and highly flammable. It poses a risk of serious damage to the eyes and may be harmful by inhalation, ingestion, or skin absorption. Wear appropriate gloves and safety goggles and use in a chemical fume hood. Keep away from heat, sparks, and open flame.

Osmium amine is highly toxic. It may be harmful by inhalation, ingestion, or skin absorption. There is a possible risk of irreversible effects. Wear appropriate gloves and safety goggles and always use in a chemical fume hood. Do not breathe the vapors.

Osmium tetroxide (osmic acid), OsO_4, is highly toxic if inhaled, ingested, or absorbed through the skin. Vapors can react with corneal tissues and cause blindness. There is a possible risk of irreversible effects. Wear appropriate gloves and safety goggles and always use in a chemical fume hood. Do not breathe the vapors.

OsO_4, *see* **Osmium tetroxide**

Ouabain is toxic to human cells by inhalation, ingestion, or skin absorption. Wear appropriate gloves and safety goggles and use in a chemical fume hood. Do not breathe the dust.

Oxygen, O_2, presents a fire and explosion hazard. The gas is heavier than air and is a strong oxidant. Vapors may cause dizziness or asphyxiation without warning. Keep away from heat, sparks, and open flame.

[32]P, *see* **Radioactive substances**

[33]P, *see* **Radioactive substances**

Packing, column, *see* **Resins**

Paclobutrazol is an irritant and may be harmful by inhalation, ingestion, or skin absorption. Wear appropriate gloves and safety glasses.

Palladium chloride causes respiratory tract irritation and may be harmful by inhalation, ingestion, or skin absorption. Wear appropriate gloves and safety glasses.

Pancreatic endonuclease I may be harmful by inhalation, ingestion, or skin absorption. Wear appropriate gloves and safety goggles.

Papain is an irritant and may be harmful by inhalation, ingestion, or skin absorption. Wear appropriate gloves and safety glasses. Do not breathe the dust.

Paraformaldehyde is highly toxic and may be fatal. It may be a carcinogen. It is readily absorbed through the skin and is extremely destructive to the skin, eyes, mucous membranes, and upper respiratory tract. Avoid breathing the dust or vapor. Wear appropriate gloves and safety glasses and use in a chemical fume hood. Keep away from heat, sparks, and open flame.

Pararosaniline chloride may be a carcinogen and is harmful by inhalation, ingestion, or skin absorption. Wear appropriate gloves and safety glasses. Do not breathe the dust or mist.

Pb, *see* **Lead**

PBL, *see* **β-Propiolactone**

Pb(NO$_3$)$_2$, *see* **Lead nitrate**

PEG, *see* **Polyethyleneglycol**

PEI, *see* **Polyethylenimine**

Penicillic acid is highly toxic and may cause cancer. It may be harmful by inhalation, ingestion, or skin absorption. Wear appropriate gloves and safety glasses. Do not breathe the dust.

Penicillin G (Procaine Salt) may cause allergic respiratory and skin reactions and may be harmful by inhalation, ingestion, or skin absorption. Wear appropriate gloves. Do not breathe the dust.

Pentafluorophenol is toxic and may be harmful by inhalation, ingestion, or skin absorption. It is also corrosive to the skin. Wear appropriate gloves and safety glasses and use in a chemical fume hood.

Pentafluorophenyl (OPFP) esters can form pentafluorophenol when they break down.

Pentafluoropropionic acid is extremely destructive to tissue of the mucous membranes and upper respiratory tract, eyes, and skin. It may be harmful by inhalation, ingestion, or skin absorption. Wear appropriate gloves and safety glasses and use in a chemical fume hood.

Pentane is irritating to the eyes, mucous membranes, and upper respiratory tract. It may be harmful by inhalation, ingestion, or skin absorption. Wear appropriate gloves and safety glasses and use in a chemical fume hood. It is highly volatile and extremely flammable. Keep away from heat, sparks, and open flame.

Pentobarbital sodium is toxic and may be harmful by inhalation, ingestion, or skin absorption. It can induce respiratory depression and sedation and presents a risk to the unborn child. Do not breathe the dust. Wear appropriate gloves and safety glasses and use in a chemical fume hood.

Pepsin may be harmful by inhalation, ingestion, or skin absorption. Wear appropriate gloves and safety glasses.

Pepstatin A may be harmful by inhalation, ingestion, or skin absorption. Wear appropriate gloves and safety glasses and use in a chemical fume hood.

Perchloric acid may be fatal by inhalation, ingestion, or skin absorption. Wear appropriate gloves and safety glasses and use only in a chemical fume hood.

Permount, *see* **Toluene**

Periodic acid is a strong oxidizer. Contact with other material may cause fire. It is also corrosive and may be harmful by inhalation, ingestion, or skin absorption. Wear appropriate gloves and safety goggles.

Petroleum ether is highly flammable and may cause central nervous system depression. It is a poison and may be harmful by inhalation, ingestion, or skin absorption. Wear appropriate gloves and safety glasses and use with appropriate ventilation. Keep away from heat, sparks, and open flame.

Phalloidin and **Phalloidin CPITC labeled, FITC labeled,** and **TRITC labeled** are extremely toxic and may be fatal by inhalation, ingestion, or skin absorption. Great care should be taken when using these compounds. Wear appropriate gloves and safety glasses and use in a chemical fume hood. Do not breathe the dust.

Phenanthroline is an irritant and is toxic if swallowed. It is harmful by inhalation, ingestion, or skin absorption. Wear appropriate gloves and safety glasses. Do not breathe the dust.

Phenazine methosulfate (PMS) is an irritant and may be harmful by inhalation, ingestion, or skin absorption. Wear appropriate gloves and safety glasses.

Phenol is extremely toxic, highly corrosive, and can cause severe burns. It may be harmful by inhalation, ingestion, or skin absorption. Wear appropriate gloves, goggles, protective clothing, and always use in a chemical fume hood. Rinse any areas of skin that come in contact with phenol with a large volume of water and wash with soap and water; do not use ethanol!

Phenol red may be harmful by inhalation, ingestion, or skin absorption. Wear appropriate gloves and safety glasses and use in a chemical fume hood.

Phenylenediamine may be harmful by inhalation, ingestion, or skin absorption. Wear appropriate gloves and safety glasses and use in a chemical fume hood.

Phenyl hydrazine (or its **hydrochloride**) is highly toxic and is a carcinogen. It may be harmful by inhalation, ingestion, or skin absorption. Wear appropriate gloves and safety glasses and use in a chemical fume hood.

Phenyl isocyanate is highly toxic, combustible, and flammable. It may be harmful by inhalation, ingestion, or skin absorption. It can cause burns. Wear appropriate gloves and safety goggles. Keep away from heat, sparks, and open flame.

Phenylmethylsulfonyl fluoride (PMSF), $C_7H_7FO_2S$, is a highly toxic cholinesterase inhibitor. It is extremely destructive to the mucous membranes of the respiratory tract, eyes, and skin. It may be fatal by inhalation, ingestion, or skin absorption. Wear appropriate gloves and safety glasses and always use in a chemical fume hood. In case of contact, immediately flush eyes or skin with copious amounts of water and discard contaminated clothing.

Phenylthiocarbamide is highly toxic and may be fatal if swallowed. It is harmful by inhalation, ingestion, or skin absorption. Wear appropriate gloves and safety goggles and use only in a chemical fume hood. Do not breathe the dust.

Phosphoric acid, H_3PO_4, is highly corrosive and is extremely destructive to the tissue of the mucous membranes and upper respiratory tract, eyes, and skin. It is harmful by inhalation, ingestion, or skin absorption. Wear appropriate gloves and safety glasses. Do not breathe the vapors.

Phosphorous pentoxide (Phosphorous oxide) is highly corrosive and causes severe irritation and burns to every area of contact. It reacts violently with water to generate heat and phosphoric acid. It is harmful by inhalation, ingestion, or skin absorption. Wear appropriate gloves and safety goggles.

Phosphortungstic acid can cause severe irritation and may be fatal. It is harmful by inhalation, ingestion, or skin absorption. Wear appropriate gloves and safety goggles. Do not breathe the dust.

Photographic fixatives and developers contain chemicals that can be harmful. Handle them with care and follow manufacturer's guidelines.

PicoGreen contains **Dimethylsulfoxide (DMSO)**. *See* **Dimethylsulfoxide (DMSO)**.

α-**Picoline** is flammable, and the effects from exposure are acute. It is harmful by inhalation, ingestion, or skin absorption. Wear appropriate gloves and safety goggles.

β-**Picoline** is flammable, and the effects from exposure are acute. It is harmful by inhalation, ingestion, or skin absorption. Wear appropriate gloves and safety goggles.

Picric acid powder (Trinitrophenol) is caustic and potentially explosive if it is dissolved and then allowed to dry out. Care must be taken to ensure that stored solutions do not dry out. Handle all concentrated acids with great care. It is also highly toxic and may be harmful by inhalation, ingestion, or skin absorption. Wear appropriate gloves and safety goggles and use in a chemical fume hood.

Piperazine is corrosive and causes burns. It is also extremely destructive to the tissue of the mucous membranes and upper respiratory tract, eyes, and skin. It is harmful by inhalation, ingestion, or skin absorption. Wear appropriate gloves and safety glasses and use only in a chemical fume hood. Do not breathe the dust.

Piperidine is highly toxic and is corrosive to the eyes, skin, respiratory tract, and gastrointestinal tract. It reacts violently with acids and oxidizing agents and may be harmful by inhalation, ingestion, or skin absorption. Do not breathe the vapors. Keep away from heat, sparks, and open flame. Wear appropriate gloves and safety glasses and use in a chemical fume hood.

Pluronic acid may be harmful by inhalation, ingestion, or skin absorption. Wear appropriate gloves and safety glasses.

PMS, *see* **Phenazine methosulfate**

PMSF, *see* **Phenylmethylsulfonyl fluoride**

PNPP, *see* **p-Nitrophenyl phosphate**

Polyacrylamide is considered to be nontoxic, but it should be treated with care because it may contain small quantities of unpolymerized material (*see* **Acrylamide**).

Polyethyleneglycol (PEG) may be harmful by inhalation, ingestion, or skin absorption. Wear appropriate gloves and safety glasses. Do not breathe the vapor.

Polyethylenimine (PEI) may cause eye and skin burns. It may be harmful by inhalation, ingestion, or skin absorption. Wear appropriate gloves and safety goggles.

Polyvinyl alcohol may be harmful by inhalation, ingestion, or skin absorption. Wear appropriate gloves and safety glasses.

Ponceau S is an irritant and may be harmful by inhalation, ingestion, or skin absorption. Wear appropriate gloves and safety glasses.

Pontamine Sky Blue, *see* **Chicago Sky Blue**

Potassium arsenate, *see* **Arsenic**

Potassium bromide, KBr, may be harmful by inhalation, ingestion, or skin absorption. Wear appropriate gloves and safety glasses and use in a chemical fume hood.

Potassium cacodylate, *see* **Cacodylate**

Potassium carbonate may be harmful by inhalation, ingestion, or skin absorption. Wear appropriate gloves and safety glasses and use in a chemical fume hood.

Potassium chloride, KCl, may be harmful by inhalation, ingestion, or skin absorption. Wear appropriate gloves and safety glasses.

Potassium cyanide, KCN, may be fatal by inhalation, ingestion, or skin absorption. Wear appropriate gloves and safety glasses and always use with extreme care in a chemical fume hood. Keep away from acids.

Potassium ferricyanide, $K_3Fe(CN)_6$, may be fatal by inhalation, ingestion, or skin absorption. Wear appropriate gloves and safety glasses and always use with extreme care in a chemical fume hood. Keep away from strong acids.

Potassium ferrocyanide, $K_4Fe(CN)_6 \cdot 3H_2O$, may be fatal by inhalation, ingestion, or skin absorption. Wear appropriate gloves and safety glasses and always use with extreme care in a chemical fume hood. Keep away from strong acids.

Potassium gluconate may be harmful by inhalation, ingestion, or skin absorption. Wear appropriate gloves and safety glasses.

Potassium hydroxide, KOH, and **KOH/methanol,** are highly toxic and may be fatal if swallowed. They may be harmful by inhalation, ingestion, or skin absorption. Solutions are corrosive and can cause severe burns. They should be handled with great care. Wear appropriate gloves and safety goggles.

Potassium iodide, KI, may be harmful by inhalation, ingestion, or skin absorption. Wear appropriate gloves and safety glasses and use in a chemical fume hood.

Potassium nitrate, KNO_3, is a strong oxidizer. Contact with other material may cause fire. It is also corrosive and may be harmful by inhalation, ingestion, or skin absorption. Wear appropriate gloves and safety goggles.

Potassium periodate may be harmful by inhalation, ingestion, or skin absorption. Wear appropriate gloves and safety glasses and use in a chemical fume hood. Do not breathe the dust.

Potassium permanganate, $KMnO_4$, is an irritant and a strong oxidant. It may form explosive mixtures when mixed with organics. Use all solutions in a chemical fume hood. Do not mix with hydrochloric acid.

Potassium thiocyanate (KSCN) causes eye and skin irritation. It may be harmful by inhalation, ingestion, or skin absorption. Wear appropriate gloves and safety goggles. Do not breathe the dust.

PPO, *see* **Diphenyloxazole**

Progesterone may be harmful by inhalation, ingestion, or skin absorption. Do not breathe the dust. Wear appropriate gloves and safety glasses and use in a chemical fume hood.

Propane is a flammable high-pressure liquid and gas and may form explosive mixtures with air. Vapor may travel a considerable distance to source of ignition and flash back. Keep away from heat, sparks, and flame. Wear appropriate gloves and safety glasses and use a mechanical exhaust.

1-Propanol, *see* **Isopropanol**

2-Propanol, *see* **Isopropanol**

*n***-Propanol,** *see* **Isopropanol**

Propidium iodide may be harmful by inhalation, ingestion, or skin absorption. It is irritating to the eyes, skin, mucous membranes, and upper respiratory tract. It is mutagenic and possibly carcinogenic. Wear appropriate gloves, safety glasses, protective clothing, and always use with extreme care in a chemical fume hood.

β-Propiolactone (PBL) is a carcinogen and mutagen and is highly toxic. It may be fatal if inhaled and is also corrosive. It may be harmful by inhalation, ingestion, or skin absorption. Wear appropriate gloves and safety goggles and use only in a chemical fume hood. Keep away from heat, sparks, and open flame. Do not breathe the dust.

Propionic acid is highly corrosive and causes burns to any area of contact. It is flammable in both liquid and vapor forms and may be harmful by inhalation, ingestion, or skin absorption. Wear appropriate gloves and safety goggles and use only with adequate ventilation. Keep away from heat, sparks, and open flame.

Propylene oxide is highly flammable, toxic, and may be carcinogenic. High concentrations are extremely destructive to the mucous membranes and upper respiratory tract. It may be harmful by inhalation, ingestion, or skin absorption. Wear appropriate gloves and safety glasses and use only in a chemical fume hood. Keep away from heat, sparks, and open flame.

n-**Propyl gallate (NPG),** *see* **Benzoic acid**

Proteinase K is an irritant and may be harmful by inhalation, ingestion, or skin absorption. Wear appropriate gloves and safety glasses.

Pseudomonas, see **Bacterial strains**

Psoralen is highly corrosive, especially to the eyes and skin, and may be carcinogenic. It may be harmful by inhalation, ingestion, or skin absorption. Wear appropriate gloves and safety goggles. Do not breathe the dust.

Puromycin is toxic and may be carcinogenic. It may be harmful by inhalation, ingestion, or skin absorption. Wear appropriate gloves and safety glasses.

Putrescine is flammable and corrosive and may be harmful by inhalation, ingestion, or skin absorption. Wear appropriate gloves and safety glasses. Keep away from heat, sparks, and open flame.

Pyridine is highly toxic and extremely destructive to the mucous membranes, upper respiratory tract, skin, and eyes. It may be harmful by inhalation, ingestion, or skin absorption. It is a possible mutagen and may cause male infertility. Keep away from heat, sparks, and open flame. Wear appropriate gloves and safety glasses and always use in a chemical fume hood.

Pyridine-*N*-oxide is toxic. Repeated exposure may affect the immune system and lungs. It may be harmful by inhalation, ingestion, or skin absorption. The effects of exposure may be delayed. It is also a strong oxidant. Wear appropriate gloves and safety glasses and always use in a chemical fume hood. Keep away from heat, sparks, and open flame.

Pyronin Y may be mutagenic. It may be harmful by inhalation, ingestion, or skin absorption. Wear appropriate gloves and safety glasses.

Quinacrine may be fatal by inhalation, ingestion, or skin absorption. Wear appropriate gloves and safety glasses and use in a chemical fume hood.

Radioactive substances: When planning an experiment that involves the use of radioactivity, include the physicochemical properties of the isotope (half-life, emission type, and energy), the chemical form of the radioactivity, its radioactive concentration (specific activity), total amount, and its chemical concentration. Order and use only as much as is really needed. Always wear appropriate gloves, lab coat, and safety goggles when handling radioactive material. **X-rays** and **gamma rays** are electromagnetic waves of very short wavelengths either generated by technical devices or emitted by radioactive materials. They may be emitted isotropically from the source or may be focused into a beam. Their potential dangers depend on the time period of exposure, the intensity experienced, and the wavelengths used. Be aware that appropriate shielding is usually of lead or other similar material. The thickness of the shielding is determined by the energy(s) of the X-rays or gamma rays. Consult the local safety office for further guidance in the appropriate use and disposal of radioactive materials. Always monitor thoroughly after using radioisotopes. A convenient calculator to perform routine radioactivity calculations can be found at:

http://graphpad.com/quickcalcs/index.cfm

Reserpine is toxic and may be carcinogenic and mutagenic. It may be harmful by inhalation, ingestion, or skin absorption. Wear appropriate gloves and safety glasses. Do not breathe the dust.

Resins are flammable and are suspected carcinogens. The unpolymerized components and dusts may cause toxic reactions, including contact allergies with long-term exposure. Avoid breathing the vapors and dusts. Wear appropriate gloves and safety goggles and always use in a chemical fume hood. Sensitivity to these chemicals may develop with repeated contact. Keep away from heat, sparks, and open flame.

Retinoic acid poses a possible risk to the unborn child. It may be harmful by inhalation, ingestion, or skin absorption. Avoid prolonged or repeated exposure. Wear appropriate gloves and safety glasses. Do not breathe the dust.

Rifampicin may be harmful by inhalation, ingestion, or skin absorption. Wear appropriate gloves and safety glasses and use in a chemical fume hood.

RNase A is an irritant and may be harmful by inhalation, ingestion, or skin absorption. Wear appropriate gloves and safety glasses. Do not breathe the dust.

^{35}S, *see* **Radioactive substances**

Salicylates may be harmful by inhalation, ingestion, or skin absorption. Wear appropriate gloves and safety glasses.

Salmonella, *see* **Bacterial strains**

SAM, *see* ***S*-Adenosyl methionine**

Saponin is an irritant and may be harmful by inhalation, ingestion, or skin absorption. Do not breathe the dust. Wear appropriate gloves and safety glasses and use in a chemical fume hood.

SDS, *see* **Sodium dodecyl sulfate**

Selenium dioxide, SeO$_2$, may be fatal by inhalation, ingestion, or skin absorption. It is highly toxic by inhalation of the dust or vapor. Wear appropriate gloves and safety glasses and always use in a chemical fume hood.

SeO$_2$, *see* **Selenium dioxide**

Serotonin may be harmful by inhalation, ingestion, or skin absorption. Wear appropriate gloves and safety glasses. Do not breathe the dust. Overexposure may cause reproductive disorders.

Sigmacote, *see* **Silane**

Silane is extremely flammable and corrosive. It may be harmful by inhalation, ingestion, or skin absorption. Keep away from heat, sparks, and open flame. The vapor is irritating to the eyes, skin, mucous membranes, and upper respiratory tract. Wear appropriate gloves and safety goggles and always use in a chemical fume hood.

Silica is an irritant and may be harmful by inhalation, ingestion, or skin absorption. Wear appropriate gloves and safety glasses. Do not breathe the dust.

Silver lactate may be harmful by inhalation, ingestion, or skin absorption. Wear appropriate gloves and safety glasses. Do not breathe the dust or mist.

Silver nitrate, AgNO$_3$, is a strong oxidizing agent and should be handled with care. It may be harmful by inhalation, ingestion, or skin absorption.

Avoid contact with skin. Wear appropriate gloves and safety glasses. It can cause explosions upon contact with other materials.

Silwet L-77 may be harmful by inhalation, ingestion, or skin absorption. Wear appropriate gloves and safety goggles. Do not inhale the aerosol.

Sinapinic acid may be harmful by inhalation, ingestion, or skin absorption. Wear appropriate gloves and safety glasses.

Sodium acetate (NaOAc), *see* **Acetic acid**

Sodium azide, NaN$_3$, is highly poisonous. It blocks the cytochrome electron transport system. Solutions containing sodium azide should be clearly marked. It may be harmful by inhalation, ingestion, or skin absorption. Wear appropriate gloves and safety goggles, and handle it with great care. Sodium azide is an oxidizing agent and should not be stored near flammable chemicals.

Sodium bicinchoninic acid, *see* **Bicinchoninic acid**

Sodium borodeuteride is a flammable solid, corrosive, and causes burns. It is water-reactive and is harmful by inhalation, ingestion, or skin absorption. Wear appropriate gloves and safety goggles and use in a chemical fume hood.

Sodium borohydride, NaBH$_4$, is corrosive and causes burns. It may be harmful by inhalation, ingestion, or skin absorption. Wear appropriate gloves and safety goggles and use in a chemical fume hood.

Sodium cacodylate may be carcinogenic and contains arsenic. It is highly toxic and may be fatal by inhalation, ingestion, or skin absorption. It also may cause harm to the unborn child. Effects of contact or inhalation may be delayed. Do not breathe the dust. Wear appropriate gloves and safety goggles and use only in a chemical fume hood. *See also* **Cacodylate**.

Sodium carbonate, Na$_2$CO$_3$, may be harmful by inhalation, ingestion, or skin absorption. Wear appropriate gloves and safety glasses and use in a chemical fume hood.

Sodium citrate, *see* **Citric acid**

Sodium cyanoborohydride is highly toxic. It is a poison, corrosive, flammable, and reacts with water. It may be fatal by inhalation, ingestion, or skin absorption. Wear appropriate gloves and safety goggles and use in a chemical fume hood. Do not breathe the dust. Keep away from water, heat, sparks, and open flame.

Sodium deoxycholate is irritating to mucous membranes and the respiratory tract and may be harmful by inhalation, ingestion, or skin absorption. Wear appropriate gloves and safety glasses when handling the powder. Do not breathe the dust.

Sodium deoxycholic acid, *see* **Sodium deoxycholate**

Sodium deuteroxide causes burns and may be harmful by inhalation, ingestion, or skin absorption. Wear appropriate gloves and safety glasses. Do not breathe the vapor.

Sodium dodecyl sulfate (SDS) is toxic, an irritant, and poses a risk of severe damage to the eyes. It may be harmful by inhalation, ingestion, or skin absorption. Wear appropriate gloves and safety goggles. Do not breathe the dust.

Sodium ethylmercurithiosalicylate, *see* **Thimerosal**

Sodium fluoride, NaF, is highly toxic and causes severe irritation. It may be fatal by inhalation, ingestion, or skin absorption. Wear appropriate gloves and safety glasses and use only in a chemical fume hood.

Sodium hydroxide, NaOH, and **solutions containing NaOH,** are highly toxic and caustic and should be handled with great care. Wear appropriate gloves and a face mask. All other concentrated bases should be handled in a similar manner.

Sodium hypochlorite, NaOCl, *see* **Bleach**

Sodium iodide, NaI, may be harmful by inhalation, ingestion, or skin absorption. Wear appropriate gloves and safety glasses and use in a chemical fume hood.

Sodium isothiocyanate, *see* **Sodium thiocyanate (NaSCN)**

Sodium *N*-lauroylsarcosinate may be harmful by inhalation, ingestion, or skin absorption. Wear appropriate gloves and safety glasses. Do not breathe the dust.

Sodium metabisulfate may be harmful by inhalation, ingestion, or skin absorption. Wear appropriate gloves and safety glasses and use in a chemical fume hood.

Sodium metabisulfite, $Na_2S_2O_5$, may be harmful by inhalation, ingestion, or skin absorption. Wear appropriate gloves and safety glasses and use in a chemical fume hood.

Sodium molybdate dihydrate may be harmful by inhalation, ingestion, or skin absorption. Wear gloves and safety glasses and use in a chemical fume hood.

Sodium nitrate, $NaNO_3$, may be harmful by inhalation, ingestion, or skin absorption. Wear appropriate gloves and safety glasses and use in a chemical fume hood.

Sodium nitrite, $NaNO_2$, is irritating to the eyes, mucous membranes, upper respiratory tract, and skin. It may be harmful by inhalation, ingestion, or skin absorption. Wear appropriate gloves and safety glasses and always use in a chemical fume hood. Keep away from acids.

Sodium orthovanadate, Na_3VO_4, may be harmful by inhalation, ingestion, or skin absorption. Wear appropriate gloves and safety glasses and use in a chemical fume hood.

Sodium perchlorate, $NaClO_4$, may be harmful by inhalation, ingestion, or skin absorption. Material is irritating to mucous membranes, upper respiratory tract, and eyes. Wear appropriate gloves and safety glasses and use in a chemical fume hood.

Sodium periodate, $NaIO_4$, and **Sodium periodate meta** are strong oxidizers and may be harmful by inhalation, ingestion, or skin absorption. They cause eye, skin, mucous membrane, and respiratory tract irritation. Wear appropriate gloves and safety glasses and use in a chemical fume hood.

Sodium pyrophosphate is an irritant and may be harmful by inhalation, ingestion, or skin absorption. Wear appropriate gloves and safety glasses. Do not breathe the dust.

Sodium salicylate is an irritant and may be harmful by inhalation, ingestion, or skin absorption. Wear appropriate gloves and safety glasses. Do not breathe the dust.

Sodium selenite may be fatal by inhalation, ingestion, or skin absorption. It is irritating to the mucous membranes, respiratory tract, eyes, and skin and may also be carcinogenic. Wear appropriate gloves and safety glasses and always use in a chemical fume hood.

Sodium sulfate, Na_2SO_4, may be harmful by inhalation, ingestion, or skin absorption. Wear appropriate gloves and safety glasses and use in a chemical fume hood.

Sodium thiocyanate (NaSCN) is an irritant and may be harmful by inhalation, ingestion, or skin absorption. Wear appropriate gloves and safety glasses.

Sodium toluene sulfonate, *see* **Toluenesulfonic acid**

Sodium vanadate is toxic and is harmful by inhalation, ingestion, or skin absorption. Wear appropriate gloves and safety glasses and use in a chemical fume hood. Do not breathe the dust. Avoid prolonged or repeated exposure.

Solvents should be handled with great care. Wear appropriate gloves and a face protector and use in a chemical fume hood.

Spermidine may be corrosive and cause severe eye and skin burns. It may be harmful by inhalation, ingestion, or skin absorption. Effects may be delayed. Wear appropriate gloves and safety goggles and use in a chemical fume hood.

Spermine may be corrosive and cause severe eye and skin burns. It may be harmful by inhalation, ingestion, or skin absorption. Effects may be delayed. Wear appropriate gloves and safety goggles and use in a chemical fume hood.

Spurr's resin contains chemicals that are carcinogens and are toxic. It may be harmful by inhalation, ingestion, or skin absorption. Wear gloves and safety glasses and use in a chemical fume hood.

Stains: Follow manufacturer's safety guidelines

***Staphylococcus aureus* (*S. aureus*),** *see* **Bacterial strains**

***Streptomyces*,** *see* **Bacterial strains**

Streptomycin is toxic and a suspected carcinogen and mutagen. It may cause allergic reactions. It may be harmful by inhalation, ingestion, or skin absorption. Wear appropriate gloves and safety glasses.

Succinic anhydride is a possible mutagen and is a severe eye irritant. It may be harmful by inhalation, ingestion, or skin absorption. Wear appropriate gloves and safety glasses and use only in a chemical fume hood. Do not breathe the dust.

Sulfo-GMBS (*N*-[gamma-Maleimidobutyryloxy]sulfo succinimide ester) may cause eye irritation and may be harmful by inhalation, ingestion, or skin absorption. Wear appropriate gloves and safety goggles.

Sulfo-MBS (*m*-Maleimidobenzoyl sulfo succinimide ester) may cause eye irritation and may be harmful by inhalation, ingestion, or skin absorption. Wear appropriate gloves and safety goggles.

Sulfosalicylic acid (dihydrate) is extremely destructive to the mucous membranes and respiratory system. Do not breathe the dust. Wear appropriate gloves and safety glasses and use only in a chemical fume hood.

Sulfuric acid, H_2SO_4, is highly toxic and extremely destructive to tissue of the mucous membranes and upper respiratory tract, eyes, and skin. It causes burns, and contact with other materials (e.g., paper) may cause fire. Wear appropriate gloves, safety glasses, and lab coat, and use in a chemical fume hood.

Surfasil is flammable and corrosive. It may be harmful by inhalation, ingestion, and skin contact and can cause severe burns of the skin and eyes. Wear appropriate gloves and safety goggles. Keep away from heat, sparks, and open flame.

SYBR GREEN I/GOLD is supplied by the manufacturer as a 10,000-fold concentrate in DMSO which transports chemicals across the skin and other tissues. Wear appropriate gloves and safety glasses and decontaminate according to your safety office guidelines. *See* **DMSO.**

SYPRO Orange/Red/Ruby contains **DMSO**, *See* **DMSO.**

TAE buffer contains **Tris** and **Glacial acetic acid.**

TAME, *see* *N*-α-**Tosyl-L-arginine methyl ester hydrochloride**

Tannic acid is an irritant. Large amounts may cause liver and kidney damage. It may be harmful by inhalation, ingestion, or skin absorption. Wear appropriate gloves and safety glasses. Do not breathe the dust.

TBE, *see* **Tetrabromoethane**

TBP, *see* **Tributylphosphine**

TCA, *see* **Trichloroacetic acid**

TCEP, *see* **Tris-(carboxyethyl)phosphine hydrochloride**

TE buffer contains **Tris-acetate.**

TEAC, *see* **Tetraethylammonium chloride**

TEMED, *see N,N,N´,N´-***Tetramethylethylenediamine**

Tergitol (Nonylphenoxy polyethoxy ethanol) is extremely irritating to the eyes and can cause permanent damage. It may be harmful by inhalation, ingestion, or skin absorption. Wear appropriate gloves and safety goggles.

*tert-***Amyl alcohol,** *see* **Amyl alcohol**

*tert-***Butanol,** *see* **Butyl alcohol-***tert*

TESPA, *see* **Silane**

Tetrabromoethane may be fatal by inhalation, ingestion, or skin absorption. It may affect liver, kidneys, and central nervous system. Wear appropriate gloves and safety goggles and use in a chemical fume hood.

Tetrabutylammonium thiocyanate may be harmful by inhalation, ingestion, or skin absorption. Wear appropriate gloves and safety glasses.

Tetrachloroauric acid may be harmful by inhalation, ingestion, or skin absorption, and is irritating to the eyes, respiratory system, and skin. Wear appropriate gloves and safety glasses and use in a chemical fume hood.

Tetracycline may be harmful by inhalation, ingestion, or skin absorption. Wear appropriate gloves and safety glasses and use in a chemical fume hood.

Tetraethylammonium chloride (TEAC) may cause allergic skin reaction and may be harmful by inhalation, ingestion, or skin absorption. It is irritating to the mucous membranes and upper respiratory tract. Do not breathe the dust. Wear appropriate gloves and safety glasses.

Tetrahydrochloride may be harmful by inhalation, ingestion, or skin absorption. Wear appropriate gloves and safety glasses.

(*N,N,N´,N´*-Tetrakis(2-pyridlmethyl)ethylenediamine (TPEN) may be harmful by inhalation, ingestion, or skin absorption. Wear appropriate gloves and safety glasses.

3,3´,5,5´-Tetramethylbenzidine, *see* **Benzidine**

N,N,N´,N´-**Tetramethylethylenediamine (TEMED)** is highly caustic to the eyes and mucous membranes and may be harmful by inhalation, ingestion, or skin absorption. Wear appropriate gloves and tightly sealed safety goggles.

Tetramethylrhodamine isothiocyanate (TRITC) may be harmful by inhalation, ingestion, or skin absorption. Wear appropriate gloves and safety glasses.

Tetrasodium pyrophosphate, *see* **Sodium pyrophosphate**

Tetrodotoxin is one of the most toxic substances known to man. Death can occur within 30 minutes. It is extremely harmful by inhalation, ingestion, or skin absorption. Wear appropriate gloves and safety goggles and use in a chemical fume hood. Do not breathe the dust.

TFA, *see* **Trifluoroacetic acid**

Thapsigargin is an irritant and may be harmful by inhalation, ingestion, or skin absorption. Do not breathe the dust. Wear appropriate gloves and safety goggles.

Thimerosal is highly toxic and harmful by inhalation, ingestion, or skin absorption. Do not breathe the dust. Wear appropriate gloves and safety glasses.

Thioanisole is an irritant and may be harmful by inhalation, ingestion, or skin absorption. Wear appropriate gloves, face mask, and safety glasses, and use only in a chemical fume hood. Do not breathe the vapors or fumes, which are toxic.

Thionyl chlorine reacts violently with water (liberates toxic gas) and causes severe burns. It is highly toxic and harmful by inhalation, ingestion, or skin absorption. Wear appropriate gloves and safety glasses and use in a chemical fume hood. Do not breathe the vapor.

Thiourea may be carcinogenic and may be harmful by inhalation, ingestion, or skin absorption. Wear appropriate gloves and safety glasses and use in a chemical fume hood.

Timentin (injectable) is an irritant and may produce allergic reactions. Wear appropriate gloves and safety glasses.

Tissues (human), *see* **Blood (human) and blood products**

TMAO, *see* **Trimethylamine N-oxide**

Toluene, $C_6H_5CH_3$, vapors are irritating to the eyes, skin, mucous membranes, and upper respiratory tract. Toluene can exert harmful effects by inhalation, ingestion, or skin absorption. Do not inhale the vapors. Wear appropriate gloves and safety glasses and use in a chemical fume hood. Toluene is extremely flammable. Keep away from heat, sparks, and open flame.

Toluenesulfonic acid is very corrosive, causes burns, and is extremely destructive to the upper respiratory tract. It may be harmful by inhalation, ingestion, or skin absorption. Wear appropriate gloves and safety glasses and use only in a chemical fume hood.

N-α-**Tosyl-L-arginine methyl ester hydrochloride (TAME)** may be harmful by inhalation, ingestion, or skin absorption. Wear appropriate gloves and safety glasses and use only in a chemical fume hood.

N-α-**Tosyl-L-lysyl chloromethyl ketone** may be harmful by inhalation, ingestion, or skin absorption. Wear appropriate gloves and safety glasses and use only in a chemical fume hood.

N-**Tosyl-L-phenylalanine chloromethyl ketone (TPCK)** may be harmful by inhalation, ingestion, or skin absorption. Wear appropriate gloves and safety glasses and use only in a chemical fume hood.

TOTO-3 iodide contains **Dimethylsulfoxide (DMSO).** *See* **Dimethylsulfoxide**

TPCK, *see N*-Tosyl-L-phenylalanine chloromethyl ketone

TPEN, *see N,N,N´,N´*-Tetrakis(2-pyridylmethyl)ethylenediamine

Transilluminator, *see* **UV light**

Tresyl chloride (2,2,2-Trifluoroethanesulfonyl chloride) is corrosive and may be harmful by inhalation, ingestion, or skin absorption. Do not breathe the vapors. Wear appropriate gloves and safety glasses and use in a chemical fume hood. Keep away from heat, sparks, and open flame.

2,2,2-Tribromoethanol may be harmful by inhalation, ingestion, or skin absorption. The vapor is also irritating to the eyes, mucous membranes, and upper respiratory tract. Wear appropriate gloves and safety glasses.

2,2,2-Tribromoethyl alcohol may be harmful by inhalation, ingestion, or skin absorption. The vapor is also irritating to the eyes, mucous membranes, and upper respiratory tract. Wear appropriate gloves and safety glasses.

Tributylphosphine (TBP) may be fatal. It may be harmful by inhalation, ingestion, or skin absorption. Wear appropriate gloves and safety goggles and use in a chemical fume hood. Do not breathe the vapor.

Tributyltin is toxic and harmful by inhalation, ingestion, or skin absorption. It may cause serious damage to the eyes and is extremely destructive to the respiratory system, eyes, and skin. It may cause burns. Wear appropriate gloves and safety goggles. Do not breathe the dust.

Tricaine methane sulfonate is an irritant and may be harmful by inhalation, ingestion, or skin absorption. Wear appropriate gloves and safety glasses.

Trichloroacetic acid (TCA) is highly caustic. Wear appropriate gloves and safety goggles.

Trichlorofluoroethane, *see* **Trichlorotrifluoroethane**

Trichlorotrifluoroethane may be harmful by inhalation, ingestion, or skin absorption. Wear appropriate gloves and safety goggles and use in a chemical fume hood. Keep away from heat, sparks, and open flame.

Triethanolamine may be harmful by inhalation, ingestion, or skin absorption. Wear appropriate gloves and safety glasses and use only in a chemical fume hood.

Triethoxysilane is flammable and highly toxic. It is extremely destructive to the mucous membranes, upper respiratory tract, eyes, and skin. It may be harmful by inhalation, ingestion, or skin absorption. Wear appropriate gloves and safety glasses and use in a chemical fume hood. Keep away from heat, sparks, and open flame.

3-(Triethoxysil)butanal, *see* **Triethoxysilane**

3-(Triethoxysil)propyl aldehyde is an irritant and may be harmful by inhalation, ingestion, or skin absorption. Wear appropriate gloves and safety glasses. Do not breathe the vapors. keep away from heat, sparks, and open flame.

Triethylamine is highly toxic and flammable. It is extremely corrosive to the mucous membranes, upper respiratory tract, eyes, and skin. It may be harmful by inhalation, ingestion, or skin absorption. Wear appropriate gloves and safety glasses and use in a chemical fume hood. Keep away from heat, sparks, and open flame.

Trifluoroacetic acid (TFA) (concentrated) may be harmful by inhalation, ingestion, or skin absorption. Concentrated acids must be handled with great care. Decomposition causes toxic fumes. Wear appropriate gloves and a face mask and use in a chemical fume hood.

2,2,2-Trifluoroethanesulfonyl chloride, *see* **Tresyl chloride**

Trihydrochloride may be harmful by inhalation, ingestion, or skin absorption. Wear appropriate gloves and safety glasses.

Triisobutylsilane may be harmful by inhalation, ingestion, or skin absorption. Wear appropriate gloves and safety glasses. Do not breathe the dust.

Trimethoprim may be harmful by inhalation, ingestion, or skin absorption. Wear appropriate gloves and safety glasses and use in a chemical fume hood.

Trimethylamine N-oxide (TMAO) causes eye irritation and may be harmful by inhalation, ingestion, or skin absorption. Wear appropriate gloves and safety goggles.

Trinitrophenol, *see* **Picric acid**

Trioctylamine may be harmful by inhalation, ingestion, or skin absorption. Wear appropriate gloves and safety glasses and use in a chemical fume hood.

Tris may be harmful by inhalation, ingestion, or skin absorption. Wear appropriate gloves and safety glasses.

Tris (carboxyethyl) phosphine hydrochloride (TCEP) is corrosive to the mucous membranes, upper respiratory tract, eyes, and skin, and can cause burns. It may be harmful by inhalation, ingestion, or skin absorption. Wear appropriate gloves and safety glasses and use in a chemical fume hood. Do not breathe the vapor or mist.

Tri-Sil is highly flammable both in liquid and vapor forms. Keep away from heat, sparks, and open flame. It may be harmful by inhalation, ingestion, or skin absorption. Wear appropriate gloves and safety glasses.

Trisodium citrate, *see* **Citric acid**

TRITC, *see* **Tetramethylrhodamine isothiocyanate**

Triton X-100 causes severe eye irritation and burns. It may be harmful by inhalation, ingestion, or skin absorption. Wear appropriate gloves and safety goggles. Do not breathe the vapor.

Triton X-114 causes severe eye irritation and burns. It may be harmful by inhalation, ingestion, or skin absorption. Wear appropriate gloves and safety goggles. Do not breathe the vapor.

Trizol may be fatal if absorbed through the skin, inhaled, or swallowed. It can also cause severe burns. Wear appropriate gloves, safety goggles, protective clothing, and always use in a chemical fume hood. Rinse any areas of skin that come in contact with trizol with a large volume of water and wash with soap and water; do not use ethanol!

Trypan blue may be a carcinogen and may be harmful by inhalation, ingestion, or skin absorption. Do not breathe the dust. Wear appropriate gloves and safety glasses.

Trypsin may cause an allergic respiratory reaction. It may be harmful by inhalation, ingestion, or skin absorption. Do not breathe the dust. Wear appropriate gloves and safety goggles. Use with adequate ventilation.

Tryptophan may be harmful by inhalation, ingestion, or skin absorption. Wear appropriate gloves and safety glasses.

Ubiquitin is an irritant and may be harmful by inhalation, ingestion, or skin absorption. Wear appropriate gloves and safety glasses.

Uranyl acetate is toxic if inhaled, ingested, or absorbed through the skin. Wear appropriate gloves and safety glasses and use in a chemical fume hood.

Urea may be harmful by inhalation, ingestion, or skin absorption. Wear appropriate gloves and safety glasses.

Urethane is a mutagen and suspected carcinogen. It is also highly toxic and is readily absorbed through the skin. It is harmful by inhalation, ingestion, or skin absorption. Wear appropriate gloves and safety glasses. Do not breathe the dust and use only in a chemical fume hood.

UV light and/or **UV radiation** is dangerous and can damage the retina. Never look at an unshielded UV light source with naked eyes. Examples of UV light sources that are common in the laboratory include hand-held lamps and transilluminators. View only through a filter or safety glasses that absorb harmful wavelengths. UV radiation is also mutagenic and carcinogenic. To minimize exposure, make sure that the UV light source is adequately shielded. Wear protective appropriate gloves when holding materials under the UV light source.

Valine may be harmful by inhalation, ingestion, or skin absorption. Wear appropriate gloves and safety glasses.

VCD, *see* **Vinylcyclohexene dioxide**

Vinblastine sulfate is an irritant and may be harmful by inhalation, ingestion, or skin absorption. Wear appropriate gloves and safety glasses. Do not breathe the dust.

Vinylcyclohexene dioxide (VCD) is irritating to the eyes, mucous membranes, and upper respiratory tract and is a carcinogen. Wear appropriate gloves and safety glasses and use in a chemical fume hood.

4-Vinylpyridine may be fatal by inhalation, ingestion, or skin absorption. Wear appropriate gloves and safety goggles. Use only in a chemical fume hood. Do not breathe the vapors.

V8 Protease is an irritant and may be harmful by inhalation, ingestion, or skin absorption. Wear appropriate gloves and safety glasses. Do not breathe the dust.

Viruses may pose special hazards, and it is recommended that all researchers working with them consult their local safety office and the handbook edited by J.Y. Richmond and R.W. McKinney. 1999. *Biosafety in microbiological and biomedical laboratories (BMBL)*, 4th edition. U.S. Department of Health and Human Services, Centers for Disease Control at:

http://www.cdc.gov/od/ohs/biosfty/biosfty.htm

X-gal may be toxic to the eyes and skin. Observe general cautions when handling the powder. Note that stock solutions of X-gal are prepared in DMF, an organic solvent. For details, *see N,N*-**Dimethylformamide (DMF)**. *See also* **5-Bromo-4-chloro-3-β-D-galactopyranoside (BCIG)**.

X-Gluc, *see* **5-Bromo-4-chloro-3-indolyl-β-D-glucuronide**

X-rays, *see* **Radioactive substances**

Xylazine may be harmful by inhalation, ingestion, or skin absorption. Wear appropriate gloves and safety glasses.

Xylene is flammable and may be narcotic at high concentrations. It may be harmful by inhalation, ingestion, or skin absorption. Wear appropriate gloves and safety glasses and use only in a chemical fume hood. Keep away from heat, sparks, and open flame.

Xylene cyanol, *see* **Xylene**

Zinc chloride, $ZnCl_2$, is corrosive and poses possible risk to the unborn child. It may be harmful by inhalation, ingestion, or skin absorption. Wear appropriate gloves and safety glasses. Do not breathe the dust.

Zinc sulfate, $ZnSO_4$, may be harmful by inhalation, ingestion, or skin absorption. Wear appropriate gloves and safety glasses.

$ZnSO_4$, *see* **Zinc sulfate**

ZnCl₂, *see* **Zinc chloride**

Zymolyase may be harmful by inhalation, ingestion, or skin absorption. Wear appropriate gloves and safety glasses.

Appendix 1

Incompatible Chemicals

The following is a partial listing of incompatible chemicals that should be kept segregated. It is intended for general information only. A more complete listing may be found in various laboratory and chemical reference manuals. Always consult your local safety office for more detailed guidelines.

Partial List of Incompatible Chemicals (Reactive Hazards)

Substances in the left-hand column should be stored and handled so that they cannot accidentally contact corresponding substances in the right-hand column.

Acetic acid	Chromic acid, nitric acid, peroxides, permanganates
Acetic anhydride	Hydroxyl-containing compounds such as ethylene glycol, perchloric acid
Acetone	Concentrated nitric and sulfuric acid mixtures, hydrogen peroxide
Acetylene	Chlorine, bromine, copper, silver, fluorine, mercury
Alkali and alkaline earth metals, such as sodium, potassium, lithium, magnesium, calcium, powdered aluminum	Carbon dioxide, carbon tetrachloride, other chlorinated hydrocarbons (also prohibit the use of water, foam, and dry chemical extinguishers on fires involving these metals—dry sand should be employed)
Ammonia (anhydrous)	Mercury, chlorine, calcium hypochlorite, iodine, bromine, hydrogen fluoride
Ammonium nitrate	Acids, metal powders, flammable liquids, chlorates, nitrites, sulfur, finely divided organics, combustibles
Aniline	Nitric acid, hydrogen peroxide
Bromine	Ammonia, acetylene, butadiene, butane, other petroleum gases, sodium carbide, turpentine, benzene, finely divided metals
Calcium oxide	Water
Carbon, activated	Calcium hypochlorite, other oxidants
Chlorates	Ammonium salts, acids, metal powders, sulfur, finely divided organics, combustibles
Chromic acid and chromium trioxide	Acetic acid, naphthalene, camphor, glycerol, turpentine, alcohol, other flammable liquids
Chlorine	Ammonia, acetylene, butadiene, butane, other petroleum gases, hydrogen, sodium carbide, turpentine, benzene, finely divided metals

Chlorine dioxide	Ammonia, methane, phosphine, hydrogen sulfide
Copper	Acetylene, hydrogen peroxide
Fluorine	Isolate from everything
Hydrazine	Hydrogen peroxide, nitric acid, any other oxidant
Hydrocarbons (benzene, butane, propane, gasoline, turpentine, etc.)	Fluorine, chlorine, bromine, chromic acid, peroxides
Hydrocyanic acid	Nitric acid, alkalis
Hydrofluoric acid (anhydrous)	Ammonia (aqueous or anhydrous)
Hydrogen fluoride	
Hydrogen peroxide	Copper, chromium, iron, most metals or their salts, any flammable liquid, combustible materials, aniline, nitromethane
Hydrogen sulfide	Fuming nitric acid, oxidizing gases
Iodine	Acetylene, ammonia (anhydrous or aqueous)
Mercury	Acetylene, fulminic acid,[a] ammonia
Nitric acid (concentrated)	Acetic acid, acetone, alcohol, aniline, chromic acid, hydrocyanic acid, hydrogen sulfide, flammable liquids, flammable gases, nitratable substances
Nitroparaffins	Inorganic bases, amines
Oxalic acid	Silver and mercury and their salts
Oxygen	Oils, grease, hydrogen, flammable liquids, solids, gases
Perchloric acid	Acetic anhydride, bismuth and its alloys, alcohol, paper, wood, grease, oils (all organics)
Peroxides, organic	Acids (organic or mineral) (also avoid friction, store cold)
Phosphorus (white)	Air, oxygen
Phosphorus pentoxide	Alcohols, strong bases, water
Potassium chlorate	Acids (see also chlorates)
Potassium perchlorate	Acids (see also perchloric acid)
Potassium permanganate	Glycerol, ethylene glycol, benzaldehyde, sulfuric acid
Silver and silver salts	Acetylene, oxalic acid, tartaric acid, fulminic acid,[a] ammonium compounds
Sodium	See alkali metals (above)
Sodium nitrite	Ammonium nitrate and other ammonium salts
Sodium peroxide	Any oxidizable substance, such as ethanol, methanol, glacial acetic acid, acetic anhydride, benzaldehyde, carbon disulfide, glycerol, ethylene glycol, ethyl acetate, methyl acetate, furfural
Sulfuric acid	Chlorates, perchlorates, permanganates

[a]Produced in nitric acid–ethanol mixtures.
(Reprinted, with permission, from *Prudent practices in the laboratory*. 1995. National Academy Press, Washington, D.C.)

Appendix 2

Most Commonly Used Isotopes

Most Commonly Used Isotopes with Their Radioactive Half-lives, Detection Methods, and Shielding Methods

Isotope	Half-life	Detection	Shielding
^{14}C	5,730 years	Geiger-Müller detector	1 cm Plexiglas
^{3}H	12.4 years	liquid scintillation	none required
^{125}I	59.6 days	scintillation detector	lead
^{131}I	8.1 days	liquid scintillation	lead
^{32}P	14.3 days	Geiger-Müller detector	1 cm Plexiglas
^{33}P	25.4 days	Geiger-Müller detector	1 cm Plexiglas
^{35}S	87.43 days	Geiger-Müller detector	1 cm Plexiglas

From the half-life, the remaining intensity can be calculated using

$$I(t) = I(t_0)\, e^{-\frac{t}{t_{1/2}} \cdot \ln 2}$$

where

$I(t)$ = intensity at the time point given (e.g., in cpm)

$I(t_0)$ = intensity at a reference date

t = time period (in days) elapsed between the time point given and the reference date

$t_{1/2}$ = half-life as given in the table above

$\ln 2$ = ca. 0.7

e = ca. 2.7

For example:

In July, $[^{32}P]$dCTP with a specific activity of 3000 Ci/mmole (reference date: July 20th) was bought. Five microliters were left in the vial after immediate use of the majority. Can these 5 µl be used after Labor Day weekend (September 6th)?

The parameters are therefore:

$I(t)$ unknown
$I(t_0)$ 3000 Ci/mmole
t 48 days (passed between July 20th and September 6th)
$t_{1/2}$ 14.3 days in the case of ^{32}P

therefore:

$I(t) = 286$ Ci/mmole

After approximately three half-lives, the specific activity is reduced about tenfold. Depending on the planned experiment, the radioactive material might still be usable.

Appendix 3

Resistance of Some Plastics to Chemicals and Sterilization

Chemical Resistance Classification

E 30 days of constant exposure causes no damage. Plastic may even tolerate it for years.

G Little or no damage after 30 days of constant exposure to the reagent.

F Some effect after 7 days of constant exposure to the reagent. Depending on the plastic, the effect may be crazing, cracking, loss of strength, or discoloration. Solvents may cause softening, swelling, and permeation losses with LDPE, HDPE, PP, PPCO, and PMP. The solvent effects on these five resins are normally reversible; the part will usually return to its normal condition after evaporation.

N Not recommended for continuous use. Immediate damage may occur. Depending on the plastic, the effect will be a more severe crazing, cracking, loss of strength, discoloration, deformation, dissolution, or permeation loss.

Resin Codes

ECTFE	Halar* ECTFE (ethylene-chlorotrifluoroethylene copolymer)
ETFE	Tefzel† (ethylene-tetrafluoroethylene)
FEP	Teflon† FEP (fluorinated ethylene propylene)
FLPE	fluorinated high-density polyethylene
HDPE	high-density polyethylene
LDPE	low-density polyethylene
PC	polycarbonate
PETG	polyethylene terephthalate copolyester
PFA	Teflon† PFA (perfluoroalkoxy)
PMMA	polymethyl methacrylate (acrylic)
PMP	polymethylpentene ("TPX")
PP	polypropylene
PPCO	polypropylene copolymer
PPO	polyphenylene oxide
PS	polystyrene
PSF	polysulfone
PUR	polyurethane
PVC	polyviinyl chloride
PVDF	polyvinylidene fluoride
ResMer	ResMer™ Manufacturing Technology
SAN	styrene acrylonitrile
TFE	Teflon† TFE (tetrafluoroethylene)
TPE	thermoplastic elastomer
XLPE	cross-linked high-density polyethylene

*Halar is a registered trademark of Solvey Solexis.
†Or equivalent.
Tefzel and Teflon are registered tradmarks of DuPont.

Chemical Resistance Summary

Classes of Substances at 20°C	ECTFE/ETFE	FEP/TFE/PFA	FLPE	HDPE	LDPE	PC	PETG	PMMA	PMP	PP/PPCO	PS	PSF	PUR	PVC (BOTTLE)	FLEXIBLE PVC TUBING	PVDF	ResMer″	TPE***
Acids, dilute or weak	E	E	E	E	E	E	G	G	E	E	E	E	F	E	G	E	E	G
Acids,** strong and concentrated	E	E	G	G	G	N	N	N	E	G	F	G	N	G	F	E	G	F
Alcohols, aliphatic	E	E	E	E	E	G	G	N	E	E	G	G	N	G	F	E	E	E
Aldehydes	E	E	G	G	G	F	G	F	G	G	F	F	N	G	N	G	G	G
Bases/Alkali	E	E	F	E	E	N	N	F	E	E	E	E	F	E	F	G	E	F
Esters	G	E	G	G	G	N	F	N	E	G	N	N	N	N	N	G	F	N
Hydrocarbons, aliphatic	E	E	E	G	F	G	G	G	G	G	F	G	G	G	F	E	G	E
Hydrocarbons, aromatic	G	E	E	N	N	N	N	N	N	N	N	N	N	N	N	E	F	N
Hydrocarbons, halogenated	G	E	G	N	N	N	N	N	N	N	N	N	N	N	N	F	F	F
Ketones, aromatic	G	E	G	N	N	N	N	N	F	N	N	N	N	F	N	F	F	N
Oxidizing Agents, strong	E	E	F	F	F	F	F	F	N	G	F	G	G	N	G	F	G	N

** Except for oxidizing acids.
*** TPE gaskets.

(Adapted, with permission, from the Nalgene Labware 2005–2006 catalog, Nalge Nunc International.)

Appendix 4

Chemical- and Liquid-Resistant Gloves

No gloves are available to protect laboratory personnel against all potential chemical exposures. Selection of the proper type of glove to protect hands from exposure to hazardous material is essential. The type of chemical being used is the determining factor in glove selection. For chemical resistance, permeation, and degradation of the glove, material selection is critical. Select the glove with the highest resistance rating and other glove properties that best suit your application. The chart below is not definitive and does not contain information on every chemical. Contact the manufacturer if you need information and data on a chemical that is not listed here.

Frequently used gloves are listed below.

1. **Asbestos gloves.** Prohibited. Kevlar and Zetex are substitutes. Refer to paragraph 5 below.

2. **Aluminized gloves.** Offer both reflective and insulating protection. The insert shall NOT be made of asbestos. Kevlar and Zetex are satisfactory asbestos substitutes. These gloves are generally used for welding, furnace, and foundry work.

3. **Coated fabric gloves.** Normally made from cotton flannel with napping on one side. The unnapped side is coated with a plastic material. This type of glove is a general-purpose protector offering slip-resistant qualities. They are used in laboratory tasks and for handling bricks and wire rope.

4. **Chemical- and liquid-resistant gloves.** Made from rubber (latex, nitrile, or butyl) or a synthetic composition such as neoprene. Frequently used gloves are described below.

    a. **Butyl rubber gloves.** Provide protection from nitric acid, sulfuric acid, hydrofluoric acid, red fuming nitric acid, rocket fuels, and peroxide. These gloves have a high impermeability to gases, chemicals, and water vapor, and resistance to oxidation and ozone attack. They have high abrasion resistance and remain flexible at low temperatures.

b. **Natural latex or rubber gloves.** Provide protection from most water solutions of acids, alkalis, salts, and ketones. Plus, they are resistant to abrasions occurring in sandblasting, grinding, and polishing. These gloves have excellent wearing qualities, pliability, and comfort, and are a good general-purpose glove.

c. **Neoprene gloves.** Provide good protection from hydraulic fluids, gasoline, alcohols, organic acids, and alkalis. They have good pliability and finger dexterity, high density and tensile strength, plus high tear resistance.

d. **Nitrile rubber gloves.** Provide protection from chlorinated solvents (trichloroethylene, perchloroethylene). They are intended for jobs requiring dexterity and sensitivity, yet they stand up under mechanical use even after prolonged exposure to substances that cause other glove materials to deteriorate. They also resist abrasion, puncturing, snagging, and tearing.

5. **Substitutes for asbestos gloves.** Gloves made of the fabrics listed below are substitutes for asbestos gloves.

a. **Kevlar.** Provides protection against heat and cold. Kevlar is a synthetic material and is used by a variety of manufacturers in their gloves. Gloves made of Kevlar material are cut- and abrasion-resistant and wear well.

b. **Zetex.** Provides protection against heat and cold. It is also a synthetic material and is used by several manufacturers in their gloves. Gloves made of Zetex material are cut- and abrasion-resistant and also withstand diluted acids (except hydrofluoric, alkalis, and solvents).

Gloves: Chemical Resistance Selection Chart

Chemical	Neoprene gloves	Latex or rubber gloves	Butyl gloves	Nitrile latex gloves
*Acetaldehyde	VG	G	VG	G
Acetic acid	VG	VG	VG	VG
*Acetone	G	VG	VG	P
Ammonium hydroxide	VG	VG	VG	VG
*Amyl acetate	F	P	F	P

Gloves: Chemical Resistance Selection Chart (*continued*)

Chemical	Neoprene gloves	Latex or rubber gloves	Butyl gloves	Nitrile latex gloves
Aniline	G	F	F	P
*Benzaldehyde	F	F	G	G
*Benzene	P	P	P	F
Butyl acetate	G	F	F	P
Butyl alcohol	VG	VG	VG	VG
Carbon disulfide	F	F	F	F
*Carbon tetrachloride	F	P	P	G
Castor oil	F	P	F	VG
*Chlorobenzene	F	P	F	P
*Chloroform	G	P	P	E
Chloronaphthalene	F	P	F	F
Chromic acid (50%)	F	P	F	F
Citric acid (10%)	VG	VG	VG	VG
Cyclohexanol	G	F	G	VG
*Dibutyl phthalate	G	P	G	G
Diesel fuel	G	P	P	VG
Diisobutyl ketone	P	F	G	P
Dimethylformamide	F	F	G	G
Dioctyl phthalate	G	P	F	VG
Diaxane	VG	G	G	G
Epoxy resins, dry	VG	VG	VG	VG
*Ethyl acetate	G	F	G	F
Ethyl alcohol	VG	VG	VG	VG
*Ethyl ether	VG	G	VG	G
*Ethylene dichloride	F	P	F	P
Ethylene glycol	VG	VG	VG	VG
Formaldehyde	VG	VG	VG	VG
Formic acid	VG	VG	VG	VG
Freon 11	G	P	F	G
Freon 12	G	P	F	G
Freon 21	G	P	F	G
Freon 22	G	P	F	G
*Furfural	G	G	G	G
Gasoline, leaded	G	P	F	VG
Gasoline, unleaded	G	P	F	VG

Gloves: Chemical Resistance Selection Chart (*continued*)

Chemical	Neoprene gloves	Latex or rubber gloves	Butyl gloves	Nitrile latex gloves
Glycerin	VG	VG	VG	VG
Hexane	F	P	P	G
Hydrazine (65%)	F	G	G	G
Hydrochloric acid	VG	G	G	
Hydrofluoric acid (48%)	VG	G	G	G
Hydrogen peroxide (30%)	G	G	G	G
Hydroquinone	G	G	G	F
Isooctane	F	P	P	VG
Kerosene	VG	F	F	VG
Ketones	G	VG	VG	P
Lacquer thinners	G	F	F	P
Lactic acid (85%)	VG	VG	VG	VG
Lauric acid (36%)	VG	F	VG	VG
Lineoleic acid	VG	P	F	G
Linseed oil	VG	P	F	VG
Maleic acid	VG	VG	VG	VG
Methyl alcohol	VG	VG	VG	VG
Methylamine	F	F	G	G
Methyl bromide	G	F	G	F
*Methyl chloride	P	P	P	P
*Methyl ethyl ketone	G	G	VG	P
*Methyl isobutyl ketone	F	F	VG	P
Methyl methacrylate	G	G	VG	F
Monoethanolamine	VG	G	VG	VG
Morpholine	VG	VG	VG	G
Naphthalene	G	F	F	G
Naphthas, aliphatic	VG	F	F	VG
Naphthas, aromatic	G	P	P	G
*Nitric acid	G	F	F	F
Nitric acid, red and white fuming	P	P	P	P
*Nitromethane (95.5%)	F	P	F	F
Nitropropane (95.5%)	F	P	F	F
Octyl alcohol	VG	VG	VG	VG
Oleic acid	VG	F	G	VG

Gloves: Chemical Resistance Selection Chart (*continued*)

Chemical	Neoprene gloves	Latex or rubber gloves	Butyl gloves	Nitrile latex gloves
Oxalic acid	VG	VG	VG	VG
Palmitic acid	VG	VG	VG	VG
Perchloric acid (60%)	VG	F	G	G
Perchloroethylene	F	P	P	G
Petroleum distillates (naphtha)	G	P	P	VG
Phenol	VG	F	G	F
Phosphoric acid	VG	G	VG	VG
Potassium hydroxide	VG	VG	VG	VG
Propyl acetate	G	F	G	F
Propyl alcohol	VG	VG	VG	VG
Propyl alcohol (iso)	VG	VG	VG	VG
Sodium hydroxide	VG	VG	VG	VG
Styrene	P	P	P	F
Styrene (100%)	P	P	P	F
Sulfuric acid	G	G	G	G
Tannic acid (65%)	VG	VG	VG	VG
Tetrahydrofuran	P	F	F	F
*Toluene	F	P	P	F
Toluene diisocyanate (TDI)	F	G	G	F
*Trichloroethylene	F	F	P	G
Triethanolamine (85%)	VG	G	G	VG
Tung oil	VG	P	F	VG
Turpentine	G	F	F	VG
*Xylene	P	P	P	F

*Limited Service
VG = Very Good, G = Good, F = Fair, P = Poor (not recommended)
(Reprinted, with permission, from the U.S. Department of Energy, OSH Technical Reference at http://www.eh.doe.gov/docs/osh_tr/ch5c.html.)